ENGINEERING
FOR LAND DRAINAGE

*A MANUAL FOR LAYING OUT AND CONSTRUCT-
ING DRAINS FOR THE IMPROVEMENT
OF AGRICULTURAL LANDS.*

BY

CHARLES G. ELLIOTT, C.E.,
Mem. Am. Soc. C. E.,
Author of " Practical Farm Drainage."

Fredonia Books
Amsterdam, The Netherlands

Engineering for Land Drainage:
A Manual for Laying Out and Constructing Drains
for the Improvement of Agricultural Lands

by
Charles G. Elliott

ISBN: 1-4101-0443-5

Copyright © 2004 by Fredonia Books

Reprinted from the 1905 edition

Fredonia Books
Amsterdam, The Netherlands
http://www.fredoniabooks.com

PREFACE.

THIS brief treatise on drainage engineering is intended for the use of those who are charged with the responsibility of making plans for and executing drainage improvements. It puts the experience and practice of years into a form which will be available to others who wish to quickly acquire the principles and practice of land drainage.

Much more might be said on some of the subjects, but the busy man of to-day prefers to have the information he seeks in concise form rather than to select what he desires from a more voluminous work.

The hydraulics of drainage cannot be computed with as much accuracy as may be done in some other branches of engineering, owing to the uncertain data available and the variable conditions which must be met. For this reason formulas of less refinement than are thought essential to some other hydraulic work may be used in making drainage computations. Good judgment should always be exercised in applying data to formulas in the class of work treated of in this book. The practical adaptation of accurate and systematic methods in dealing with land drainage questions is regarded by the author as highly essential to the successful prosecution of such work.

Complicated and merely theoretical engineering is to be avoided. Simple and practical methods which are

of general application should be especially sought for
and learned.   It has been the purpose of the author to
emphasize and make clear those points which the stu-
dent, the busy agriculturist, and the practical engineer
should know.

CHARLES G. ELLIOTT.

WASHINGTON, D. C.,
November, 1902.

CONTENTS.

CHAPTER X.

SIZE OF LATERAL DRAINS.

CHAPTER XI.

OPEN DRAINS.

CHAPTER XII.

DRAINAGE OF BARNYARDS, CATTLE-LANES, ETC.

CHAPTER XIII.

ROAD DRAINAGE.

CHAPTER XIV.

DRAINAGE DISTRICTS.

CHAPTER XV.

ESTIMATES OF COST.

CHAPTER XVI.

ENGINEERING FOR LAND DRAINAGE.

CHAPTER I.

INTRODUCTION.

THE DRAINAGE ENGINEER—THE AGRICULTURIST
AND SOIL DRAINAGE.

The Drainage Engineer.

THE importance of the work of the drainage engineer
may in a measure be appreciated when we consider the
magnitude of land drainage projects which have been
executed during the century which has just closed.
The immense tracts of land in the Old World which have
been reclaimed from the inroad of river and sea and are
now the greatest food-producing lands in the world are
monuments to the skill of engineers who planned and
executed these works. None of these drainage projects
has been carried out without the aid of engineers of
high ability, as well as of great energy and originality.
He who has in this way contributed to the world's
wealth and the happiness of man is as worthy of high
esteem as he who discovers some rich island, or by his

researches ascertains the means by which a dread disease may be averted.

The civilizing effects of the drainage of these great tracts upon the people immediately concerned are difficult to measure. It is sufficient to say, however, that it has had no small part in leading men away from primitive superstitions and rude practices to higher ambitions, nobler impulses, and purer morals. It is said that when the drainage of the North Level, a part of the celebrated Fens of England, seemed assured one of the Fen poets made the following versified predictions:

> " With a change of elements suddenly
>   There shall a change of men and manners be:
>   Hearts thick and tough as hides shall feel remorse
>   And souls of sedge shall understand discourse:
>   New hands shall learn to work, forget to steal;
>   New legs shall go to church, new knees shall kneel."

Not only this, but in the development of the fertility of a large area of all grain-producing agricultural lands drainage has been extensively practised, and is justly regarded as one of the most necessary adjuncts of successful soil culture.

Land drainage in this country has had its principal growth during the last twenty years. Its practice reaches from the simple drainage of a garden costing only a few dollars to that of tracts containing thousands of acres, involving elaborate plans and the expenditure of large sums of money. It includes the drainage of all kinds of lands which require it, either for profit or for health, and hence embraces a wider range of topics than, upon first thought, seems to be involved.

The drainage engineer should be qualified for his

chosen field of work, understanding that this is an age of specialties, and that without devoting thought and investigation to his particular pursuit he cannot expect to be proficient in it. He should understand the drainage characteristics and capabilities of different soils, and treat them according to their needs, keeping in mind the ultimate use which is to be made of them. The various means of accomplishing this work in the most thorough and economical manner belong to his special field of investigation and practice. There are sound principles upon which drainage is founded, but there are no "standard plans" for accomplishing it, as there are for the construction of many other kinds of engineering work. Writers upon the subject of drainage frequently make a mistake along this line by recommending certain plans and methods of work for all localities without regard to differences of soil, climate, physical conditions, or the use to which the land will be put after it is drained. He who measures his skill in drainage by his success with one piece of work in one locality may easily flatter himself unduly, for the reason that scarcely any two projects can be planned and worked alike with equal success.

There is a difference between the work of the surveyor and that of the engineer, though both may be done by the same person. The work of gathering definite data and facts and putting them in orderly form for future use is strictly the work of the surveyor. He may do all this and not have the requisite skill to use the data thus obtained in designing and laying out an effective and economical drainage system. The surveyor secures data; the engineer plans and executes from data. One collects material; the other builds from the material collected. In making his surveys the engineer should aim to secure sufficient information to properly design the work, yet not so much that the labor and expense involved in securing it would not be justifiable from a strictly business point of view.

The drainage engineer deals with corporations, commissioners, and private individuals as their professional expert and counsellor. He should make his employer's case his own, and use his best energies to solve the difficulties which arise, having due regard to sound practice and enduring results. He should put himself in harmony with the work and constantly keep in mind the specific object which he is expected finally to reach.

He should be proficient in the art of clearly representing his data and plans by creditable drawings and concise language. His statements should be divested of all unnecessary technicalities in order that they may be plain to the common reader. This does not, of course, imply that his knowledge or experience should be meager, but that, out of the abundance of his information, he should select that part which will bear upon the case in hand, and so express it that it will be available to others.

With truth it may be said that greater proficiency is needed in the practice of drainage engineering than at present exists. The work has not been systematized as has other professional work, nor investigated with that thoroughness which its importance demands. Much trouble arises both from incompetence of engineers and over-competence—if the expression may be allowed—of their employers. As before intimated, the engineer must, if possible, look through the eyes of his employer in order to get his view of the matter and appreciate the difficulties as they appear to him. If after investigation the two differ in opinion and the employer is unwilling to accept the plans and methods of the engineer without modifications which seem to him unwise, he should, at least, put himself on record as not endorsing the objectionable features.

Above all, the drainage engineer should exercise a high order of common sense, good judgment, and honesty in the management of the work intrusted to him. He should not let his ideas of engineering nicety carry him to a point where much of the work he does will have no particular value, and yet he should seek to do his work in a professional way and with accuracy. A well-balanced enthusiasm should characterize him in all of his work, and by fairness in dealing he should secure the confidence of all with whom he has professional or business relations.

The Agriculturist and Soil Drainage.

The soil is the farmer's business capital. He has exchanged a certain sum of money for it, or come into possession of it by inheritance, and must now look to its products for returns. He has before him for solution the various problems connected with soil culture and its relation to profit and loss. The soil becomes a receptacle for his money and a field for intelligent labor. Good husbandry strenuously insists on a thorough preparation of the seed bed and an intelligent after-cultivation of the plant. It also demands a wise

and economic use of the products of the soil, be they
grain, forage, or fruits. The end to be sought in all
of this is that the farmer may receive a profit over all
and still have his capital, the soil, intact and unim-
paired.

One of the well-recognized means of bringing about
this gratifying result—making the farm pay—is to remove the surplus moisture from the soil. By surplus is
meant more than is needed. The surplus moisture
should be removed *through* the soil, if possible, rather
than *over* it. The drainage of the soil is by no means
an innovation, nor is it a work remaining yet in an experimental stage, except as regards a better understanding of its application to various soils, and for the
purpose of demonstrating the scientific changes which
result from practical work along this line.

The growth of the drainage improvement for agricultural purposes has always been governed by the en-
vironment of the farmer. While he had sufficient land
which, in its natural condition, was drained and required only the ordinary and primitive methods of cultivation, there was no occasion for adding to his arable
estate those lands which would require more than the
ordinary expenditure of money and labor. That time
is past. Now every progressive farmer looks upon
each unimproved acre of land as an item on his farm
expense account. For does he not pay taxes upon
it in common with his most productive land? Does
it not cost him as much to cultivate as does the adjoining field of rich and friable loam? Is it not a blot
upon an otherwise fair rural picture, to say nothing of
the financial features which with the farmer and business

man are more weighty? While this is true of the small farmer who delights in high-grade and, sometimes, artistic agriculture, there are large tracts of fertile land lacking only suitable drainage to fit them for the most profitable cultivation. The enterprising agriculturist, be his interests large or small, will find an ample field for the exercise of such knowledge as he may choose to acquire in the theory and practice of land drainage. Nor should he fail to avail himself of all opportunities presented for mastering this useful subject. The fact that our national government and our State experiment stations are devoting marked attention to such subjects as "The Movement of Ground-water," "Moisture and Crop Distribution," "Mechanics of Soil Moisture," "Moisture Determination in Soils," indicates that these subjects and their bearing upon soil drainage are worthy of the most careful study by practical men.

In the discussion of topics pertaining to general farming, fruit-growing or stock-raising as given in the agricultural papers or at the sessions of Farmers' Institutes, the subject of soil drainage takes a prominent place, and the value of such an improvement of the soil is emphasized and its advantages enlarged upon by all progressive farmers. Like every other operation in which the management of the soil is concerned there is a right way and a wrong way of performing it: there may be a good way and a better way; there is only one best way. Some soils require no artificial drainage, some a little, and some a great deal, in order to yield the best results. Good judgment and some knowledge are required to adapt the method of improvement to the land to be treated. The landowner should perform the work well without unnecessary cost. He should know how the work should be done, or be able to employ some one with sufficient skill and professional honesty to do the work for him.   There is no mystery connected with the theory and practice of land drainage, as some would have us believe, neither is an instinct born in men which will relieve them from the necessity of acquiring knowledge of this work in the old-time way.

Drainage practice must be adapted to the needs of each tract of land to be improved.   In other words, each farm, and in many cases each field, presents a special problem in drainage.   How much more diverse, then, must be the practice in different States and latitudes?   The land-drainer should not think that a method which has proved efficient in one locality will necessarily be the best in another, or that when he has successfully drained one tract or farm he can drain any other by the same plan.   Yet the principles are the same for all; it is only the application of them that varies.

Some ultra-practical men discourage the farmer from attempting to learn the history and progress of land drainage, urging that the present state of the practice is sufficient.   Life is not too short for a man to avail himself of the experience of others along any line of work in which he is interested, and he acts wisely when he seeks to acquire all information pertaining to his business.   It is not proposed, however, in these pages to give such a history but to outline in plain language the best practice at the present time.

The subject will be treated from an engineering

standpoint, but that need not deter the farmer from becoming conversant with the entire subject, nor from studying it in a scientific way. It is a subject which has merited the careful attention of intelligent land-owners and of eminent hydraulic engineers.

The profits accruing from the drainage of fertile land are of two kinds: first, the increased yield of grasses, cereals, and fruits which has a direct money value to the producer; second, the increased healthfulness of the community where drainage has reclaimed all of the waste land. This latter has a money value which is difficult to measure. The following example clearly illustrates both of the above statements:

The Indiana Bureau of Statistics made an investigation of the influence of tile-drainage on health and crops, selecting a single township in the State where drainage was one of the marked improvements, and taking a period of five years before drainage began, and five years after most of the township had been tile-drained. By consulting farmers who lived in the township during both periods, they found that the average crop of wheat in the five years before drainage was $9\frac{1}{2}$ bushels per acre. The same land after drainage for five consecutive years produced an average of $19\frac{1}{4}$ bushels per acre. The average yield of corn in the first five years was $31\frac{1}{4}$ bushels per acre. In the five years after drainage the average yield was $74\frac{1}{4}$ bushels per acre. The physicians who answered calls in the township were requested to report from their books, and it was found that in the first five years there had been 1480 cases of malarial disease. In the second period there had been only 490 cases of malarial disease.   With such facts before us it will require no argument to convince the average citizen that drainage has largely increased the health and wealth of that community, and thereby added materially to the prosperity of the State.

This subject involves the conservation of soil moisture as well as the removal of surplus water, the care of private and public roads, the sanitary drainage of the home and grounds, subjects which receive too little thought from the average landowner.   One who chooses to devote a small fraction of his time to acquiring a knowledge of the principles and best practice along these lines, and to apply the same in the light of his own observation and common sense, may become sufficiently proficient to direct the operations on his own estate, in fact become his own engineer if he cares to engage in that business.

The writer appreciates the many practical difficulties which present themselves to the landowner who personally attends to all of the detail work connected with the improvement of his farm and home, as well as those that must be met by the investor who improves his holdings for the purpose of deriving a larger annual revenue from the capital invested.   Many drainage projects necessitate cooperation on the part of interested landowners, without which private interests suffer, and will continue to suffer, until certain men who oppose the combined effort become persuaded that it will be to their profit to join the movement under equitable provisions of the law, bear their proportion of the expense, and receive the benefit accruing from the work. For this reason as well as others heretofore mentioned, landowners should be conversant with the principles, benefits, and best practice of land drainage, not neglecting to acquire information regarding the legal drainage rights of adjoining landowners and the general principles of cooperative work.

to supply the needs of man only partially under-

CHAPTER II.

SOILS.

THE soil, one of the essentials to the existence and well-being of the human race, is one of the most complex products of nature. With all the acquirements of which man can boast he cannot create a pound of soil, understand the intricacies of its composition, nor yet avail himself fully of the wealth locked up within this most familiar of all natural objects. Nature has apparently brought out the choicest selections from her storehouse and placed them at the service of man in the form commonly known as soil. Its varieties are unnumbered, its capabilities unmeasured, and its adaptability to supply the needs of man only partially understood.

Origin of Soils.—Soils are broken and decomposed rocks. Before the external forces of nature acted upon them they were as barren and useless as the clean peaks of the mountains or the washed rock in the bed of the stream. Many of the changes to which the original rocks were subjected in bringing them into the new and useful combination are unknown, but the general process can be quite accurately understood from examples which may be witnessed in nature by any interested observer.

The mosses which are found on the rocks are a low

12

order of vegetation subsisting upon elements contained in the air and in the crude rock. Not many years elapse until the moss thickens to such an extent that it appears to be growing in a thin bed of soil, while directly beneath it the rock is decomposed and portions of it drop away in scales as soon as the covering is removed. This is accomplished by the combined action of heat moisture and the changes incident to climatic influences. The mosses continue to grow, bits of decomposed rock drop and lodge at the base, forming a shelving bed of soil which readily supports vegetable growth. Rains possibly wash it away to some low valley, there to be mingled with decomposed rocks from other localities, all of which in time build the fertile soil. In it may be found numberless kinds of rocks, and as vegetation increases portions of organic matter mingle with the material and contribute the humus so valuable for the production of certain crops. It seems incredible that the hardest of rocks known should succumb to the action of climate and be reduced to a condition of impalpable fineness, yet a little observation on the part of the investigator will confirm the statement. The reader has doubtless observed a tree growing upon an apparently dry rock. A nearer view discloses crevices which the roots of the tree have penetrated. The crevices are lined with rock in all stages of disintegration which increase year by year with the growth of the tree. Further than this, the force of the wedge which the root forms splits off a ledge of rock which in turn disintegrates and adds to the volume of forming soil.

A ledge of red shale disintegrates rapidly under

climatic changes, and the result is a belt of red clay soil on a bench of the hillside below. In such cases all stages of soil formation may be seen from the rock-like shale to the plastic clay. The lava-covered slopes of volcanic mountains become covered with soil in the same manner. Time and the changes incident to a humid climate will convert lava, at one time a seething mass, into a fruitful soil.

One peculiarity of these natural changes is that the rocks by being brought to such a state of comminution lose their identity. The minute particles of different rocks become so thoroughly blended that their original condition can only be inferred by fragments of unchanged rock which may chance to be found in the mixture. The hundreds of forms known to the geologist and mineralogist under distinctive names, each one of which when chemically analyzed is found to contain from six to twelve elements, will never be known after being changed to soil. Hence it is only the general processes which relate to the origin of soils that are of particular interest in the discussion of this subject. The composition of soils can be known only through chemical analyses, and, since the same element is found in a variety of rocks, it is impossible to determine from such analyses what the original rocks were. A physical analysis will determine the comparative fineness of the particles constituting the soil, a matter of much importance in its treatment.

Sedentary Soils are those which remain where they are formed and constitute a covering for the rock from which they originated. They have usually little depth and comprise but a small part of useful soils.

Transported Soils are those which have been removed from their original rock beds by the action of glaciers, floods of water, or by streams which have carried soil particles in suspension and deposited them as sediment. In this way decomposed rocks from widely distant localities have become mixed together in an inseparable mass. One form of transported soils is known as the *drift* which originated during the Glacial Epoch, a period when the present surface of the country was covered to a great depth with fields of ice. This kind of soil is usually distinguished by rounded rocks of various sizes called boulders, and by fragments of rocks whose edges have been rounded by friction, all of which are incorporated with the soil proper. As might be correctly inferred, the varieties of soil found in the area affected by glacial action include every possible shade of difference. The moving glacier from whose melting mass rocks and clumps of soil were constantly being deposited, and subsequently ground by the passing mountain of ice, formed one of the later geological epochs and one which is of great interest and importance to the northern part of the United States and Canada and as far west as western Iowa.

Alluvial Soils consist of worn and rounded materials which have been transported by the agency of moving water and deposited as sediment. The possible conditions under which soils can be formed in this way are without number. Alluvial deposits have been formed in all periods of the world's history. Water trickling down a granite slope carries forward the kaolinite arising from decomposition of feldspar, and the first hollow gradually fills up with a bed of clay. In valleys are thus deposited the gravel, sand, and rock dust detached from the slopes of the neighboring mountains. Lakes and gulfs become filled with silt brought into them by streams. Alluvium is found below as well as above the drift, and recent alluvium in the drift region is very often composed of drift material rearranged by water currents.

Organic Matter of the Soil.—As before noted, vegetation plays an important part in the conversion of rocks into soil. The lower orders of plants, such as the lichens and mosses, prepare the way for grasses and forests. The decay of vegetation adds to the soil a brown or black friable substance commonly known as humus which gives off gases and aids in the further conversion of inert material into productive soil. The gases such as carbonic acid and ammonia are largely held in the soil, their volume depending upon the quantity of vegetable matter which the soil contains, and the supervention of warm wet weather.

The proportion of carbonic acid contained in the pores of different kinds of soil compared with that found in the ordinary atmosphere is strikingly shown in the following analysis made by Boussingault and Lewy.

CARBONIC ACID IN 10,000 PARTS OF AIR BY WEIGHT.

Ordinary atmosphere.....................	6
Air from sandy subsoil of forest............	38
" " loamy " " "	124
" " surface soil " "	130
Air from surface soil of vineyard...........	146
" " " " old asparagus bed...	122
" " " " newly manured land.	233
" " " " pasture land	270
" " " rich in humus	543

Kinds of Soils.—Soils are known to the agriculturist by names drawn from their external appearance or from some peculiarity which they show when worked. Many of such names have a local application only.

Clayey Soils are commonly characterized by extreme fineness of texture and by great retentive power for water. When subjected to a mechanical analysis their particles are found to be the finest of all soil particles.

Sandy Soils are those which contain 80 per cent or more of sand. Silica or grains of quartz withstand the disintegrating agencies beyond all others, and hence when once in the soil never change their form. However, there are all kinds of sandy soils, from the one which contains but little to the soil carrying 90 per cent of sand.

Gravelly Soils contain an abundance of small stones or pebbles which in themselves are worthless, but aid in a mechanical way to keep the soil open, assist in drainage, and store solar heat. Many gravelly soils are exceedingly fertile.

Loamy Soils are those intermediate in character between sandy and clayey. They can be worked freely, not having sufficient clay to be heavy, nor sand and gravel in such large quantities as to make them too open.

To express suggested differences we have the terms *Sandy Loam, Sandy Clay, Gravelly Loam*, and *Gravelly Clay*.

Gumbo Soils are loams with sufficient plastic clay mixed with them to make them exceedingly sticky or adhesive when wet. They are fertile soils when properly cultivated.   In some instances a layer of gumbo is found beneath a fine bed of loam, and supports it as a subsoil.

Muck is a black soil composed largely of vegetable matter and is found in swamps.   It frequently requires exposure to the atmosphere for a time before it can be treated as a workable soil.

Peat is partially decayed swamp turf which when dry will burn readily.   The underlying bed is usually muck or blue-black clay.

Hard-pan is the name applied to a tough, impenetrable layer underlying a fairly fertile soil.   A hard-pan proper is made up of soil particles which are being cemented together again by the solutions of lime, iron, or silicates that descend through the soil.   Commonly speaking, however, any hard clay subsoil is termed hard-pan.

Each Soil an Individual.

Each soil possesses a composition and character of its own, and it follows that its capabilities, requirements, and treatment should be taken up individually. Soil investigations are necessarily experimental, be they made with reference to productive capabilities, drainage properties or irrigation possibilities.   Each case should be taken up by itself and studied with special reference to its character and condition.   When plants are chemically analyzed they are found to contain elements found in the soil with the exception of aluminum.   However, the element which appears most abundant in the plant is frequently found in the most meager quantities in the soil which produces it, suggesting the fact now well understood that it is not so much the quantity of a needed element in the soil which is required as its availability to plants. The conditions induced by proper tilth moisture, and the use of such fertilizers as will by chemical combinations release and put in proper condition for plant assimilation the otherwise inert elements of the soil, are matters determined by experiment.

Drainage Properties of Soils and Subsoils.

Soil is the surface land that is cultivated and which produces vegetable growth. In general terms, it is the surface stratum of earth.

Subsoil is the stratum of earth upon which the soil rests.

The dividing line between the two is not clearly marked as a rule, so that the terms are usually understood to apply respectively to that depth of surface land which may be cultivated, and the layer immediately beneath. Soils and subsoils are of almost every conceivable color, composition, and physical structure. Their treatment for the growth of plants and for the support of important engineering and architectural structures, and their investigation respecting their relation to human diseases, engaged the attention of men of thought for ages. Men and animals move upon them, plants derive their nutriment from them, and water is stored among their particles. The soil is a great laboratory wherein is developed and from which is dispensed the Creator's supplies for man's temporal wants.

With respect to their drainage we speak of them as *open* and *retentive*, the terms expressing, not their power of retaining certain quantities of water, but the readiness or rapidity with which water moves among the particles when a means of drainage is offered. Between the very open soils and the very retentive ones there are numberless degrees of difference which must be expressed by qualifying terms if they are properly described.

This may be illustrated by a few examples which will serve to show the bearing of these terms. A soil or subsoil composed largely of sand or gravel offers but slight resistance to the movement of water among the particles, so that a single drain as an outlet relieves the soil of drainage-water for a very considerable distance from it. There are instances of drainage districts in which the opening of a single deep channel has drained to a considerable degree of thoroughness certain lands lying a mile each side of it. There are other localities where the effect of such drain does not extend one hundred feet, and still others where forty feet is the limit. Upon these natural conditions of the soil largely depend the means that should be employed to properly drain it.

The relation of the soil to the subsoil should also be carefully observed. The subsoil may be compact and retentive, while the soil directly above it may be open, or the opposite conditions may exist. The subsoil may not be parallel to the surface in its general conformation, but a retentive clay may crop out near the surface at some points, and recede in others, thus forming pockets or basins underneath the surface where water is retained, much to the injury of the surface soil. These frequently occur upon the tops of hills, or upon hillside slopes where the surface indications do not lead one to suspect such conditions. Fig. 1 represents a section of soil of this character.

FIG. 1.—Section showing Effect of Clay Subsoil upon Natural Drainage.

The characteristics of soils as indicated above are general, but are sufficient to point out what should be looked for when their drainage is undertaken. These general qualities are, however, dependent upon many minute differences, such as mechanical fineness and physical structure of the particles, the attraction which their surface has for water films, the chemical composition of the component parts, and many other essential particulars which have been observed and are still being investigated by the soil physicist.

Water of the Soil.

Water which affects the soil exists in two conditions.

Hydrostatic Water is visible to the eye and free to obey the laws of gravity. It is water which is found between the particles of the soil and passes off through drains either natural or artificial. It is frequently spoken of as *drainage-water*.

Capillary Water is that which is held within the fine pores of the soil by the surface attraction of its particles.   It is commonly called *moisture* and is the water which is left in the soil after it is drained.

Speaking in general terms, about 50 per cent of the volume of a soil is empty space, that is, it contains only air and water.   The results of experiments hereafter given show that the volume of empty space ranges from 37 per cent as found in sandy soils to 65 per cent in soils composed largely of clay.   This space is so divided up by the very large number of grains of soil that the spaces between the grains are extremely small.   When a soil is only slightly moist the water clings to the soil grains in a thin film.   The force which holds water to the grain of soil is called surface tension.   The water is called capillary water.   It is like a soap-bubble with a grain of sand or clay inside instead of air.   Where the grains come together the films are united into a continuous film of water throughout the soil.   If more water enters the soil, the film thickens and there is less exposed water surface.   If the empty space is completely filled with water, gravity alone will act with its greatest force.   If the soil is nearly dry, there will be a great deal of this exposed water surface, a great amount of surface tension, and hence gravity will have no effect.   In other words, there will be no hydrostatic water, since the force of surface tension is able to retain the entire supply of water about the particles of which the soil is composed.

Hydrostatic water moves through the soil with greater or less rapidity and freedom, according to the resistance which the soil particles offer to its passage. It moves upward only when forced to do so.   It moves downward and laterally in obedience to the laws of gravity. Capillary water moves in any direction in accordance with the laws of capillary attraction which exists between a liquid and a solid when brought into contact.

The movement of hydrostatic and capillary water through the soil is necessary to its healthy condition as far as the growth of plants is concerned. Capillary water moves to supply the demand of evaporation, plant-roots, and other portions of the soil which are deficient in moisture. Water moves downward from the surface by gravity and supplies needed capillary water which is held by the surface tension of the soil particles. The remainder passes off as hydrostatic water through drains either natural or artificial, or, suitable drainage not being provided, it remains to retard the growth of plants and in other ways injure the soil.

Water of the Soil that is Used by Plants.

Plants take their nutriment from the soil by means of rootlets which grow in close contact with soil particles. All plant-food taken through these rootlets must be in solution and in the condition of capillary water or, as it is commonly called, moisture. In nature's preparation of plant-food mineral matter must be dissolved, organic matter be decomposed, gases absorbed by water, and the whole stored away in liquid form within the minute pores of the soil, there to be seized upon by the rootlets of plants and thence appropriated to their growth.

It matters not how much fertility there may be present in the soil, if it is not put in this form it is not available to plants. Hydrostatic water is useful only in replenishing the supply of capillary water. When this is accomplished, the surplus should be drawn off by the process of drainage in order that air and gases may take its place. These in connection with heat and moisture aid in the decomposition of organic and mineral components of the soil which constitute its fertility. Excess of water retards, and in many cases altogether prevents this process.

It is the province of drainage to remove the surplus water and retain the capillary water, since it is in the latter form only that water is appropriated by useful plants. By the removal of hydrostatic water the chemical forces of the soil are permitted to work freely in their laboratory, and thus prepare the elements of the soil for plant use.

Experimental Investigations.

Many of the changes which take place in soils by reason of the presence of water in varying quantities are now being made the subject of careful examination at some of our experiment stations. This is particularly true of mechanical analyses of soils, and observations upon the movement of water in soils possessing different physical characteristics. It is interesting to note that the operations and effects of drains upon the soil as heretofore recorded by close observers, agree very closely with the results of experiments made in recent years for the purpose of ascertaining the facts with scientific precision. To the soil physicist, the engineer, and the observing agriculturist these investigations are in every way interesting and useful. While experiments along this line are by no means complete, yet much work has been done which bears directly upon the theory and practice of land drainage, some of the results of which will be briefly outlined in this chapter.

Soil Structure.

Soils and subsoils are composed of solid grains of variable sizes which touch each other at certain points on their surfaces. The comparative size of soil grains varies greatly, and is an important and interesting subject for investigation.

According to a mechanical analysis made in 1891 at the Maryland Experiment Station, the diameter of soil grains for the several materials named is as follows:

Gravel................	2	to 1	millimeters
Medium sand........	1	to .5	"
Fine sand...........	.25	to .1	"
Very fine sand......	.1	to .05	"
Silt................	.05	to .01	"
Fine silt...........	.01	to .005	"
Clay...............	.005	to .0001	"

Note.—Millimeter = .03937 in.

This is only comparative, as by the method used for analysis the absolute size could not be obtained, but a reliable method for classifying soils with reference to the size of soil grains was hit upon and used.

The empty space between these grains has also been determined for a variety of soils, the following being the result of some of the experiments:

Sandy truck soil................ 37.29 per cent
Wheat land.................... 42.72 " "
Barren clay .... 47.19 " "
Gummy land.................. 61.54 " "
Pipe-clay.................. ...... 65.12 " "

By comparing the fineness of the grains with the volume of empty space, it is seen that the finer the grains the greater the volume of empty space a soil contains. That is, a clay holds or is capable of holding 27.8 per cent of volume of water more than the sandy soil noted, while the grains of the clay are about five hundred times smaller than those of the sandy soil. By comparing the results of several analyses of soils and subsoils, it is found that the results which appear in the tables referred to, characterize all soils.

These investigations show that clays are made up of the smallest grains, and contain the greatest volume of empty space, and from the results of other experiments it appears that the volume of empty space of soils in general is in direct proportion to the per cent of clay which they contain. When this space becomes charged with water, the rate of movement under a given head, as would be readily and correctly inferred, is more rapid through a sandy soil than through one composed of a considerable per cent of clay. While the sandy soil contains less total volume of space and hence will hold a smaller volume of water than the clay soil, the spaces are larger individually than those in the clay and hence less resistance is offered to the flow of water through them.

Experiments were made at the Station with a num-

ber of different soils with reference to this point. Tubes
were filled with the varieties of earth to be investigated,
and water passed through the soil under a constant free
head of two inches above the soil. The time which,
under these conditions, one inch of water passed through
three inches of soil was noted. The following are a
few of the results obtained: A sandy soil containing
3.9 per cent of clay, time $5\frac{1}{2}$ minutes. A heavy red
clay with 28.8 per cent of pure clay, time 133 minutes.
A black gummy land with large grains of sand and 26
per cent of clay, 16 minutes.

These experiments bring out the facts regarding the
mechanical structure of soils from which may be drawn
some very helpful conclusions. It must be kept in
mind, however, that natural soil conditions are often
different from those existing in the portion of soil which
has been experimented with. The ground beneath its
surface contains seams and channels made by the entrance and decay of plant-roots, or is changed by the
separation of joint clays into little cubes, or possibly
the surface has been puddled by some means, thereby
essentially changing the structure in such a way as to
give us quite different results. The outlet for water is
not always at hand as provided for in experiments.
The percolation of water through the surface soil is
affected by cultivation and by the application of fertilizers, the latter having the effect of changing the
arrangement of the soil grains and making the soil
more retentive of moisture.

Capillary Movement of Water.

During the year 1897 very satisfactory experiments were conducted by the Division of Soils of the U. S. Department of Agriculture for the purpose of determining the quantity of moisture present in soils at different depths by means of the electrical method. This method is based upon the changing electrical resistance between two plates which are buried in the soil, and consists in measuring the resistance by means of a mechanism devised for the purpose. The study of the results obtained by these experiments in connection with the rainfall record and the drainage of the soil, will well repay the agriculturist or engineer who is interested in soil study.

The table on p. 29, taken from Bulletin No. 12, U. S. Department of Agriculture, Division of Soils, is interesting and useful in showing the per cent of water carried by a soil at Washington, D. C., under three different conditions: the first, growing wheat; the second, cultivated bare, and the third, well mulched.

From other experiments noted in the same Bulletin the following are selected for the purpose of showing the per cent of clay and corresponding per cent of water carried by the several soils noted, during the growing season:

At Newbern, N. C., on the best type of early truck land the soil to a depth of three or four feet has about 2.8 per cent of clay, and averages about 8.5 per cent of water for the crop season.

MOISTURE RECORD AS DEDUCED FROM ELECTRICAL RESISTANCE OF SOILS AT GROUNDS OF U. S. DEPARTMENT OF AGRICULTURE, WASHINGTON, D. C.

Day of M'nth (1897)	August				September				October			
	Wheat	Bare	Mulched	Rain	Wheat	Bare	Mulched	Rain	Wheat	Bare	Mulched	Rain
	Per ct	Per ct	Per ct	Inch	Per ct	Per ct	Per ct	Inch	Per ct	Per ct	Per ct	Inch
1	10.75	17.90	18.75		9.38	16.15	22.43		10.16	18.05	22.99	
2	19.58	17.37	18.75		9.17	16.27	22.25		9.57	17.80	22.43	
3	19.41	17.80	18.56									
4	18.26	17.64	18.65	0.03	8.73	16.24	21.90		8.90	17.34	21.66	
5	17.77	17.49	18.56						8.39	17.07	21.19	
6					8.33	16.32	21.73		7.95	17.46	20.97	
7	17.82	17.76	18.56		8.19	16.32	21.73		7.69	17.07	20.97	
8	17.54	17.52	19.18	.29	7.95	16.44	21.73		7.54	16.44	20.97	
9	20.50	19.74	21.90		7.84	16.38	21.73		7.36	16.44	20.97	
10	19.49	18.57	22.43	2.06	7.73	16.32	21.57					
11	17.77	17.88	22.25						20.50*	16.38	20.97	0.03
12	16.98	17.64	23.00		7.59	16.32	21.41		21.33	20.91	27.77	.95
13					7.46	16.38	21.57		20.88	24.20	26.28	.03
14	14.92	17.06	22.43	.03	13.22*	16.38	21.41		20.09	22.47	25.42	
15	14.02	16.38	22.43	.05	12.58	16.38	21.41		18.79	21.16	25.42	
16	13.54	16.29	22.07		11.82	16.61	21.41		17.77	20.37	24.63	
17	12.97	16.21	21.90		11.26	16.55	21.10					
18	12.74	15.10	21.90						14.13	18.51	23.93	
19	12.05	16.05	21.00		9.57	16.73	20.95	0.19	13.22	18.09	23.93	
20			21.90		8.90	16.55	20.66		21.27	20.49	30.32	.85
21	11.60	15.95			8.47	20.81	20.81		19.93	20.88	27.26	.08
22	12.45	16.83		.29	9.52	18.71	25.35		19.54	20.13	26.82	
23	12.30	16.61			23.26	18.57	24.28	.63	19.17	28.76	25.42	
24	11.82	17.20	25.13	.06	19.93	29.87	24.28	.78				
25			20.13			26.76	25.35					
26	11.21	16.32			14.72		24.28					
27	10.71	16.35			12.74	22.94	23.93					
28					11.65	21.63	23.93					
29	9.53	16.05			10.86	20.91	22.09					
30	9.52	16.21	22.25	.09		20.21						
31												

* Irrigated.

At Windsor, Conn., light sandy tobacco soil containing 1.3 per cent of clay averages 10 per cent of water for the growing season.

Heavy limestone, tobacco, corn, wheat, and grass land at Litiz, Pa., containing 36 per cent of clay, carries 23 per cent of moisture.

At Lexington, Ky., blue grass and tobacco soils which maintain about 35 per cent of clay, carry 22 per cent of moisture.

Results of experiments at Germantown, Ohio, on cigar tobacco soil which is also good for corn, wheat and grass, show $27\frac{1}{2}$ per cent of clay in the subsoil and 22 per cent of moisture.

Further experiments only confirm the above and indicate that our heavy soils suitable for grass and grain contain from 20 to 36 per cent of clay and carry from 20 to 23 per cent of moisture, while the light, sandy truck soils have as low as 3 per cent of clay and carry 8 per cent of moisture.

Compare these results with the volume of empty space in the several examples of soils, and we can determine with approximate correctness the volume of water that must be removed from a saturated soil before it becomes fitted for the several crops which it is desired to grow.

Effect of Fertilizers upon Soil Moisture.

From some experiments made by the Department it is shown that the addition of certain fertilizers placed in the soil causes it to retain moisture with greater tenacity. This is accounted for by supposing that the action of the fertilizer upon the particles of soil reduces their size, thereby making them more retentive of moisture. This, however, has not been demonstrated. It is a fact well known to every observing cultivator of the soil that good barnyard manure spread upon a portion of the field conspicuous for its dryness will result in a good showing of moisture during the dry season. If a soil is too wet, it is frequently made more so by the addition of well rotted manure; but if it be underdrained, the results will be most pleasing and profitable.

Percolation through Frozen Ground.

The impression is quite general that while the ground is frozen there can be little or no percolation through it. This is far from being true. As soon as the surface of the ground becomes slightly softened by the action of snow or rain, the shrinkage-cracks, worm-perforations, and root-courses always present in the soil, at once become available for the passage of drainage-water, and as a result underdrains may be operative in midwinter. Water is filtered through the soil and its fertilizing ingredients deposited in the soil below the frost line.

The freezing of the soil, however, is of great assistance in "fining" the surface, disintegrating the grains composing the subsoil, and, as above noted, adding fertility by the process of filtering surface water. It is readily understood that frost penetrates a drained soil deeper than a saturated one because the atmosphere takes the place of removed water. For this reason the drained soil thaws earlier in the spring. The warm surface air enters, and the soil becomes fitted for cultivation several days earlier than one which contains too much water.

Heaving of Soils.

One great benefit derived from proper drainage is that soils do not heave to such an extent as to injure plants. When soils which are saturated with water freeze, the water expands about one eighth in volume, and as a result the surface of the soil is raised, carrying with it plant-roots which are not anchored below the frost line. A thaw melts the ice, and the soil under ordinary conditions settles back to its original position, but plants with shallow rooting remain in their raised position. Alternate freezing and thawing, which take place in some localities several times during the winter, not infrequently leave the roots of clover, wheat, and rye partially or wholly out of the ground. Fields of clover and wheat have been ruined in this way, and meadows of timothy and orchard-grass have been greatly injured during one winter. If the soil is not saturated, that is, if there is air space in it and the moisture has sufficient space for expansion when it freezes without materially disarranging the soil grains, no such injurious effects as have been named will follow. The heaving effect of frost upon soils may be demonstrated in an interesting way by the following experiment. Before the ground freezes in the fall drive pegs of about one inch section and six inches long flush with the surface of the ground, some in soils well drained, and others in wet soils. At the opening of the spring many of the pegs which were driven in the wet soil will be found partly raised out of the ground, and some of them will be found lying upon the surface, as

FIG. 2.—Heaving of Wet Soils.

shown in Fig. 2. Those driven into well-drained soil will be found to have been but slightly moved by the frost.

The heaving of the soil has much to do with the unsettled condition of undrained dirt roads, for the reason that the soil grains become completely disarranged and thrown out of their natural position so that the compactness of the load-bearing surface of the road is destroyed.

Conservation of Soil Moisture.

While the removal of surplus moisture from the soil is of first importance, the retention of a proper quantity is equally necessary. The object of underdrainage is to secure and maintain a golden mean between a dry and a wet soil. All observing and thorough cultivators are particular during the growing season to give the soil a thorough surface cultivation as soon after every rain as practicable. This pulverization of the surface —the finer the better—has been found to serve the purpose of a mulch in dry weather, and to be conducive to plant growth especially in underdrained soils.

The benefit as explained by some is due to the dryness of the covering, which is secured by frequent cultivation, and which induces the formation of a new moisture line just beneath the surface. Another explanation, and the more tenable one, is this: The rainfall saturates the surface and to some extent causes the soil particles to assume a new and closer relation to each other. These when dried by the sun and wind form a thin surface crust, which increases the tension and capillarity of the soil particles so that moisture from below is carried rapidly to the surface and evaporated. Cultivation breaks up this compact condition of the soil and retards the capillary action of its surface layer. Frequent surface culture, which is now recognized as a most important adjunct to successful soil cultivation, is made possible by underdrainage, since the soil can be cultivated soon after any rain without risk of injury.

CHAPTER III.

LAND DRAINAGE PRACTICE.

Land Drainage is the removal of surplus water from the soil.

Various descriptive terms are used in connection with the word drainage according to the purpose for which this removal is effected. When done for the better growth and protection of useful plants it is called *Agricultural Drainage*. When done for the better construction and maintenance of public highways it is called *Road Drainage*. When done for the benefit of health, either in city or country, it is called *Sanitary Drainage*.

It will be the object of these pages to discuss these questions and describe the methods which engineers, or those expecting to act as such, are called upon to deal with in the planning and execution of work pertaining to land drainage.

The definition of drainage indicates that it is a very simple operation, yet its practical attainment is accompanied with no little uncertainty unless correct principles and methods are followed out. Investigations along this line are somewhat difficult, for the reason that the action of drains is hidden from the eye, and is known only by the effect which is produced. Nature has in many localities provided thorough drainage of the soil. In others it is only partly accomplished, the remainder of the work to be done by the enterprise of man.

He who drains land for any purpose whatever should put it down as a first principle that he is only aiding Nature, and hence he must work in accordance with her laws. We frequently take too much credit to ourselves in thinking that we have invented some new method, when we have only developed the plans as they have been pointed out to us by Nature. A very little artificial work in addition to natural advantage will sometimes bring about complete drainage, while in other cases great labor and expense will be necessary to accomplish results no more complete.

The first question to be asked where agricultural drainage is contemplated is, will the land be productive after drainage? If this question, after proper investigation, must be answered negatively, the matter should be dropped, if the agricultural value of the land is the only one to be considered. Some wet lands have no productive value before drainage, nor will they afterwards, though it may be said for our encouragement that such cases are exceptional.

The second question is, can the land be drained, and if so, how, and at what cost? It will be the writer's object to deal with this question in its practical details, under appropriate divisions of the subject.

Underdrainage.

No soil is completely drained of its surplus water, unless it is done by the process known as underdrainage. It may be done by the means which Nature has provided, as is the case where the subsoil contains a stratum of sand, gravel, or other permeable material, into which water from above finds its way by gravity, and thence passes to some water-course with which the drainage stratum has free communication. This gives the most complete drainage possible. When natural underdrainage is wanting, or is defective, then artificial drainage should be resorted to. Round drain-tiles are the most suitable for this purpose, and are universally acknowledged to furnish the most serviceable and complete artificial means of draining the soil now known.

Water enters the tile drain through the joints or between the ends of the tiles. Ordinarily it enters from the bottom, being brought there by gravity and held to the depth at which it may be in the tile by the lateral pressure of soil water. Water from the surface presses directly downward until it reaches a line where the soil is saturated. It then moves only as the water below the line of saturation is drawn off by the drain, when it in turn passes downward and laterally until it reaches the drain. The rapidity of this movement is measured by the hindrance offered to the water, by the nature of the soil particles, and by the ability of the drain to remove water as fast as it is brought to it. The line of saturation rises and falls as the supply of water increases or diminishes, receding during the time of least supply into the lowest plane it can occupy. At the time when the soil has just ceased to supply drainage water to the tile, a very slight rainfall, or even the change of temperature incident to night-time, will cause the drain to discharge. No water enters the tile drain until capillary water has been supplied to the soil.

The material out of which the tiles are made does not affect their efficiency as drains, nor the facility with which soil water enters the drain.

Plane of Saturation.

It will be readily understood, from the explanation just given regarding the action of a single drain upon the soil, that as the lateral distance from the drain is increased the plane of saturation rises, for the reason that the water in passing toward the drain encounters the resistance of the particles of the soil, which resistance requires a certain head to overcome, or there can be no movement of water. The angle which a line in this plane makes with a horizontal passing through the floor of the drain must vary with the degree of soil resistance. Upon this depends the lateral distance to which a single line of tile will drain the soil, and the depth at which drains should be placed. This line of saturation has been assumed from observations upon the action of drains to be a curve of some kind, but no definite investigations have been made upon the subject until recently.

In 1891 observations were made at the Wisconsin Agricultural Experiment Station in a tile-drained field bordering upon a lake, a part of which was provided with natural underdrainage alone, for the purpose of determining the actual contour of the plane of saturation. The following is a brief account of the method used and the results obtained:

" The lines of tile in the field were laid 33 feet apart and about four feet below the surface. Small wells were sunk midway between the lines of tile and were therefore $16\frac{1}{2}$ feet distant from drains on either side. The soil of the field consists of 6 or 8 inches of medium clay loam, followed by $2\frac{1}{2}$ to 3 feet of clay, below which is a stratum of rather coarse sand, in the upper surface of which the tiles are usually laid. The tiles are three inches inside diameter, and are laid on a grade of about two inches in 100 feet. At the time the levels were taken the tiles were discharging only one twentieth of their capacity.

" The observed contour of ground-water in this field at 8 A. M., May 13, forty-eight hours after a rainfall of 87 inch, is represented in Fig. 3. The highest water level in any well between these lines of tile on this date was one foot, in the case of well *A* measured above the

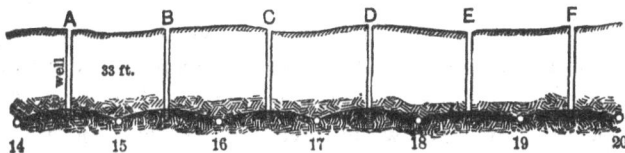

FIG. 3.—Line of Saturation between Tile Drains 48 hours after a Rainfall of $\frac{9}{10}$ of an inch.

level of drain No. 14. The least was about .3 foot, in the case of well *E* above tile No. 18. Both wells *C* and *E* were sunk into a sand containing a considerable amount of gravel, and to this fact is probably due the less steep gradient at these places. Between well *B* and tile 16 two other wells were sunk, one two feet from the drain, and the other midway between the

drain and well B.   In the well two feet from the drain the
water stood .3 foot above the top of the tile and in the
other .45 foot above.   The profile would present, therefore, a more or less curved contour, convex upward."

Prof. King, under whose direction the observations
were made, draws the following conclusions:

"Assuming the water level at the several lines of
tile to be flush with the tops of the tile, and regarding
the water surface as presenting a right line of section,
the mean gradient for the ground-water surface plane
of saturation would be 1 foot in 25.3 feet.   There were
other wells sunk outside the range of tile drains for the
purpose of ascertaining the height of the ground-water
above the water surface of the lake.   Of these he says:
"In well 29, 150 feet from the lake shore, the water
stood 7.2 feet above the level of the lake on June 27,
1892, and this would give a gradient of 1 ft. in 20.8.
In the case of the well at the Hall to which I have referred as having a water level 52 feet above the lake,
and situated about 1250 feet from the shore, the mean
gradient would be 1 in 24.04.   In the fall of 1888,
Sept. 10, when the water level in the wells could not
have been affected by lateral percolation, the gradient
between well 29 and the lake was 1 in 35.8."

He observes further that the water line of the tile-
drained field under consideration did not remain as described, but as water was carried away by the drains,
the line was drawn down at a uniform rate, falling fast-
est on the highest ground where the level was highest.

These researches serve to indicate in a tangible way
the relation of the line of saturation to underdrainage.
The method of averages used by Prof. King in drawing

his conclusions leads to an error as far as drainage is concerned. The greatest rise is found to be 1 foot in 16½ feet, and the least 1 foot in 55 feet.

Since soils in which drains are placed may be like any one of those upon which observations were made, no soil would be found to which an average of widely different results could apply except by accident or co-incidence. It follows that the plane of saturation with respect to the position of tile drains must be determined for each kind of soil which is to be drained. Sugges-tions upon this point will be made when we come to consider the application of these physical soil charac-teristics to the location of drains.

Sources of Water Affecting the Soil.

In locating drains it is necessary to consider the source of the water to be removed. Primarily the rain-fall is the source of all water-supply, but in land drain-age work it is known by various terms describing its local source.

It is known as *surface water* when the direct rainfall rests upon or flows over the surface of the land; *local soil water* when a part of the surface water sinks di-rectly into and saturates the soil and the subsoil; *flood water*, when water from some adjoining source flows upon land, thereby throwing a greater quantity of water upon it than the natural rainfall supply; *ooze or seep-age water* when it finds its way through the soil from some higher elevation, and is arrested in its course by a less slope or a change of subsoil structure (the locality where the water appears is called "spouty");

spring water when it proceeds from some distant and constant source following a channel of its own until it reaches a free outlet, where it shows its presence in a definite and constant quantity.

Systems of Tile Drains.

The various arrangements of drains according to the requirements of the surface and soil to be treated are called *systems*.

A Main Drain is one which is used to collect drainage water from smaller drains and conduct it to some open ditch or natural stream.

A Sub-Main is a drain which discharges into a main, and is itself a receiving-drain for lines of smaller tile.

A Lateral is the smallest drain in the system, and discharges into a main or sub-main.

The Natural System.—This is the kind of drainage that is first practised, and consists in laying some lines of tile in natural depressions which are particularly wet and troublesome to the agriculturist, as represented in Fig. 4. It deserves the name of system only because in many kinds of soil and localities it is all that is required to make the drainage quite complete. It is an aid to natural drainage, and in fact completes it where the higher land naturally drains itself into adjoining depressions. The occasional lines of tile then put in must carry the drainage of the natural watershed, thus compelling the drains to act as mains for a considerable area. For this reason more complaint is made of the incapacity of tile drains located in this way than where more frequent drainage is practised. The Natural

System is the skeleton which may be developed into a more complete system, if afterwards found necessary, provided the size of the drains is proportioned to the area to be finally improved.

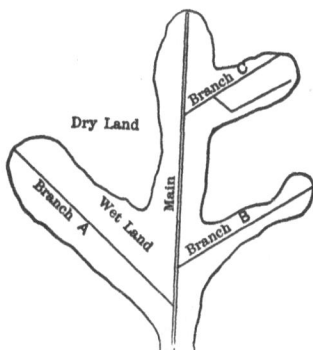

FIG. 4.—Natural System.

The Grouping System.—This may be applied to such land as has basins or sloughs, and also areas of dry land, or such as it does not seem desirable to drain. The field is divided into small drainage sections so that one outlet will serve for each division, and a main laid in the lowest land of each separate drainage basin. The drainage may be completed by lateral lines converging toward the main at such distances apart, and having such lengths, as may be adapted to the purpose. The method is shown in Fig. 5.

The Gridiron System.—This is the old and generally practised system where thorough drainage is carried out. Systematic drainage generally implies the location of parallel drains at uniform distances over the entire field. *Thorough drainage*, however, is so removing water from the entire field as to secure uniform moisture and texture in all conditions of weather. Where the soil is alike in the tenacity with which it holds moisture the system should be uniform, and every part of the ground brought under the influence

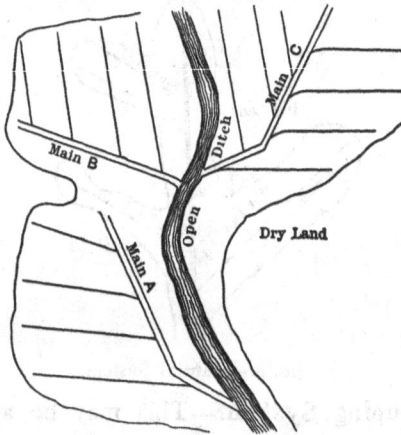

FIG. 5.—Grouping System.

of drains at regular intervals. But when the soil varies, or the surface is diversified by ponds, sloughs, and draws, thorough drainage means lines with reference to the different conditions. The gridiron system consists of equidistant parallel lines with mains and sub-mains for collecting and conducting the water to some point of exit. It is economy to have the laterals enter the mains at right angles, but for completeness and efficiency they should so enter that the currents of the two streams may coalesce and increase rather than retard the flow of the main. This system is illustrated in Fig. 6.

Double-main System.—This is applicable in broad, flat sloughs where it is desirable to use two lines of

smaller tile instead of one large main through the centre.   It is sometimes necessary to diverge the lines
toward the head, making two systems, and running

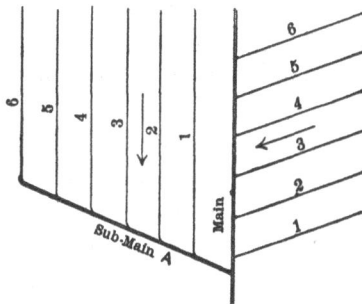

FIG. 6.—Gridiron System.

laterals into each from both sides.   In draining hillsides and wet slopes it is best to lay lateral drains down
the slope, at such intervals as are required, discharging

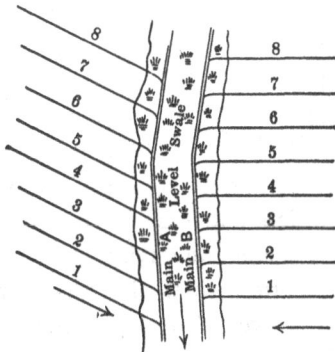

FIG. 7.—Double-main System.

into a collecting-drain.   In such cases have the collecting drain near the base of the slope, that the laterals
need not pass through a flat bottom, which would re-

tard the flow. But in locating mains in this way note that unless the slough slopes but little toward the centre line, one centre main of sufficient capacity gives better results. There are cases where this system may be followed advantageously both with respect to cost and efficiency, while in others it would prove expensive and faulty. See Fig. 7.

Single-line System.—By this is meant the plan of laying parallel lines in the direction of the greatest slope, giving each a separate outlet into an open ditch as shown in Fig. 8. This is adapted to large areas which have very slight fall and where it is consequently necessary to use every inch of it in grade and depth of

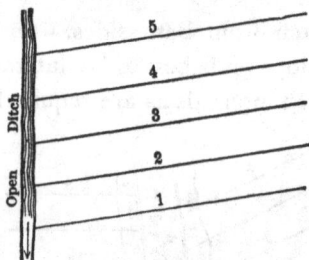

FIG. 8.--Single-line System.

drains, and also to avoid the use of excessively large tile for mains which under such circumstances would be required if a system involving mains and laterals were used. A caution which should be emphasized in locating drains on level land where grades must necessarily be very light is to avoid overcharging mains, a thing that is frequently done because the difference in grades is so often overlooked.

These different systems, or modifications of them, may be used on different tracts, and in various localities where considerations of economy and efficiency may suggest their appropriateness. It is safe to say that there is no tract of land requiring drainage to which some of them will not apply. Much will be saved in outlay and gained in efficiency by carefully adapting the system to the particular tract to be treated.

Principles to be used in Locating Drains.

Lay mains in the line of natural drainage. There are but few tracts of land that do not have some natural surface drainage, or places where the water gathers and in flood-time flows off. It is also true that, as a rule, the direction of the water of the soil is towards such places, and in order to intercept it and carry it away the main should be located there. As has been stated, we must, if possible, work in the line of natural drainage if we expect to obtain efficiency of work and economy in construction. If we consider the drainage of some distant point or tract without reference to benefiting land along the line by which it may be reached by a drain, then the question hinges upon the difference in cost of the line by way of some near cut, and the more circuitous and natural route. The shortest and straightest drain is the best provided it does the desired work as well. It is usually the case that the line of natural drainage may be straightened by short cuts here and there, making the drain less expensive and more efficient, without impairing its value as a drain in the natural course. It should be said in this connection that there are many flats, ponds, basins, etc., which can be more economically reached by a main drain through some short cut than by the natural over-flow course. This is a matter, however, that should be examined with care before a location is made.

Laterals should be laid in the line of greatest slope. Many think that by extending a drain across a slope, water coming through the soil from above will be intercepted by the drain and thus be prevented from passing further toward the foot of the slope. Practice has proved this to be a mistake. Lines for conveying the drainage-water may be located at right angles to the slopes if placed so far down on the bottom land

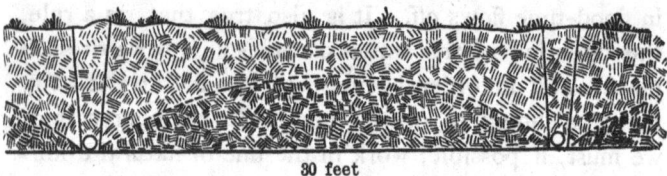

30 feet

FIG. 9.—Water-line in Retentive Clay Soils.

that the grade of the drain is greater than the slope of the surface at the side, as a few facts will show. Water oozes through the soil along the line of steepest descent, at all times seeking a lower place where it can remain at rest. If a drain is placed across this course of soil water, the descent of the soil channels being greater than that of the drain, water will flow out of the joints of the drain and continue to ooze through the soil, only a small part being conveyed away by the drain. Place the drains up and down the slope, and all water coming into the drain will be carried away quickly, and little currents induced to flow toward the drain from both sides. See Fig. 9.

While the above refers particularly to hillsides requiring drainage, it is also applicable to flat land having any slope whatever. There are sags, swales, and ponds into which an outlet tile must be extended by the most feasible course; after that the general rule applies.

Avoid short laterals where a system can be adopted in which long parallel laterals can be used. This is a matter that has to do with the economy of the work rather than with its efficiency. Every main or sub-main will, of itself, drain the land for a certain distance on either side of it. All laterals, in order to reach these mains, must be laid through the belt of land thus drained, and hence a part of each lateral will be useless except to conduct the water to its receiving drain. The fewer junctions there are in a given tract the less waste of length of lateral drains will be there. There are localities where, on account of the contour of land, the short laterals are necessary.

Make the lines as straight as practicable, and change direction by easy curves. Drains cannot always be made straight from one end to the other, yet short serpentine crooks should always be avoided. Tangents may be run and connected by good curves which will admit of the drain being put in the proper place and accomplish the work far better than can be done by irregular crooked lines which usually mark the small watercourse. The disadvantages of a crooked line are that the tiles are laid with greater difficulty and more imperfectly, there is a loss of grade where it is needed, the friction of the running stream against the walls of the drain is greater than in straight lines, and a greater length of drain will be required to accomplish the same purpose.

Bring all land which is deficient in natural drainage under the influence of tile drains. This requires the investigation of the entire watershed for the purpose of determining how complete the natural drainage is. The engineer should adopt in his own mind some standard of the degree of thoroughness with which he proposes to drain a given tract, and locate his drains with reference to the natural wetness of the land. He should find out whether the water comes from the surface of some adjoining higher land, or from distant springs, or is seep-water from slopes. If parts of the field are naturally dry, or as dry as it is proposed to make the other parts, he should pass it by and put drains in the wetter portions so as to bring them up to the standard. It may be remarked that portions of land which are supposed to have sufficient natural drainage have afterward been found deficient in this respect, when compared with the land that is thoroughly tiled.

Data Required for Locating Drains.

The knowledge of a piece of land which is necessary for the proper laying out of a drainage system may be obtained in one of two ways or partly by both.

First, the engineer may, by carefully inspecting the land with the aid of some one who is familiar with both surface and soil peculiarities, determine upon the proper system and mark out the lines readily. There is a feature connected with the location that is gratifying to the engineer, which is that, when the correct and nat-

ural plan for locating has been hit upon, the whole system may be developed with ease.

Second, more or less work with a levelling instrument
may be required in order to obtain the facts necessary
for determining upon the best plan of work. The
slopes may be so slight, or so deceptive to the eye, and
the lines of natural drainage and best points of outlet
so obscure, that it will require an instrumental survey
to determine them. This involves a certain class of
topographical work which will be described hereafter.

After the lines have been located upon the ground,
or, more properly speaking, the general plan of work
has been decided upon, then their location in the ground
as to depth and grade must be done with the level, if
any degree of accuracy in the construction of drains is
expected to be attained. This part of the work will be
described in another chapter.

CHAPTER IV.

LEVELLING AND TOPOGRAPHY.

Levelling.

Levelling is the art of finding how much one point is higher or lower than another.

A Level Line is one which is perpendicular to the direction of gravity. *A Levelling Instrument* is any instrument by which a level line can be accurately determined.

A Datum Plane, or " Datum," is the initial plane or point in the plane from which all heights or elevations are computed.

The Elevation of a point is its height when referred to datum, or its vertical distance above or below datum.

A Levelling-rod is a graduated staff for measuring the distance from the line indicated by the instrument to the point whose elevation is desired. *A Target Rod* has a sliding disc which is moved by the rodman to the position indicated by the man at the instrument. The rodman is expected to call off the reading of the figures as indicated by the position of the disc. *A Speaking-rod* is one graduated with such distinctness that it can be read by the instrument man with precision. The speaking rod is preferable for use in all drainage work.

Bench-marks are permanent objects whose elevations are determined and recorded for future reference.

Levelling Instruments.

The engineer's spirit-level in some of its more or less expensive forms is the most accurate instrument and the one with which the most rapid work can be done. Telescope-levels are sufficiently low in price, so that it is not wise to use the poor and cheap kind of levels which are recommended by many as sufficiently accurate for drainage purposes, especially where the work to be done is of any magnitude. The fact that most of our drainage work requires the utmost accuracy attainable is not appreciated by many. The care and adjustment of the instrument should be learned and can usually be done with the aid of little manuals which the instrument-makers furnish. Some of the levels have a graduated circle, by means of which horizontal angles can be measured and made use of in platting the lines along which levels are taken. For this work, however, a compass either attached or separate from the level is much to be preferred.

The rod is a necessary companion to the level. There is a great variety of forms, each one having its advocates. Some form of the speaking-rod or self-reading rod is to be preferred to any form of the target rod for the work herein described. It is capable of being used more rapidly, it makes the levelman more independent of his rodman because he takes his own readings, and in making cross-sections it can be used instead of a tape-line. The divisions and colors should be so displayed as to be clearly seen and easily read. For ease of computation, the decimal scale of feet and tenths is best, though some prefer to use feet and inches.

Fig. 10 shows a form of rod which has been found convenient and serviceable for drainage surveys, and one which can be readily and cheaply made. It is made of a strip of straight-grained white pine, 1 inch thick, 2½ inches wide, and 10 feet long. The ends are shod with ½-inch iron to protect them from battering. The rod is cut in two in the middle and a good strap hinge set in even with the face, so that the rod can be folded together for convenience in transportation. It is fastened open while in use by a rib of wood, which is screwed fast to the back, and covers the joint, and has a movable bolt and thumb-nut for use in holding the rod open or shut as desired. The accompanying cut shows the manner of graduating it. The dark spaces showing tenths of a foot are red on the rod. The foot figures are large and are painted red. The tenth figures are black, and the small squares along the centre line are also black. There are two hundredth spaces,

FIG. 10.—Self-reading Rod.

and may be divided by the eye so that the rod reads to single hundredths of a foot. The variety and combination of colors are such as to be clearly read at a distance of from three hundred to five hundred feet, according to light and power of the glass.

The chain for measuring horizontal distances is also necessary in making level results available. The 100-foot steel chain is perhaps the most convenient and serviceable for drainage surveys. It is, however, open to the objection that it is not often correct in length, and hence should be compared frequently with some standard. The band chain should be kept on hand for accurate measurements and for testing the link chain, but its liability to be broken in the hands of ordinary workmen, as well as the disadvantage of requiring two hands for setting a pin at the fore end, makes it less desirable than the link chain, from which pins are set from the ends of the handles. The link chain holds its place in weeds and grass better than the tape, can be easily thrown across a stream, and is more convenient for running in curves.

Two or more flag-poles, steel pointed at one end, and each bearing a flag of cloth half white and half red, about 8 in. by 12 in. in size are needed for marking out courses for the chainmen to follow in staking out the lines.

A small hand ax for driving the stakes and a hand basket for carrying them completes the outfit. A set of steel marking-pins will often be found convenient.

Taking Levels.

In order to make the process of levelling as simple as possible from beginning to end, and also keep the results in the best form for use, some method of procedure and of keeping notes should be adopted that will be general and apply to all cases that the drainage

engineer will have to deal with.   There is a variety of
forms for keeping level notes, as well as methods of
making drainage computations, but the methods here
described and recommended are in quite general use
among engineers and possess the merit of being simple
and adapted to all kinds of work coming within the
sphere of levelling.

The Field-book.—A good form and size for a book
in which to record and keep the level notes is one with
pages about 4 inches by 6½ inches containing about
eighty-five leaves.   Rule the left-hand page into five

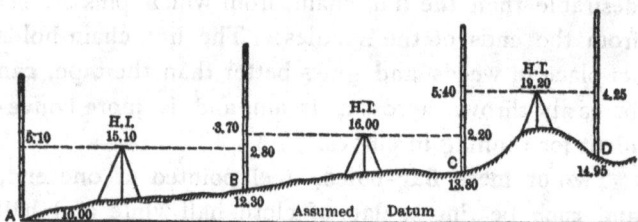

FIG. 11.—Levelling.

columns, and head them as shown in level notes accompanying Fig. 11.   The right-hand page will be
left for entering diagrams and explanatory notes.   A
size larger than this is unnecessary and will also be
found inconvenient to carry in the pocket.

Level Practice.—Select some bench-mark or other
point from which it is proposed to start, and if its elevation has not been determined before, assume one
which will be convenient to use without introducing
minus expressions.   If we begin low down on some
water-course, perhaps 10.00 will do; if higher up 20.00,
30.00, or 100.00 should be used as the elevation of the

starting-point. Call this point A and write its elevation in the elevation column opposite station A. (See
Fig. 11, and the notes to accompany the figure, both
being used to illustrate level practice.) Set the level
up midway between this point and the next point B
whose elevation is to be obtained. Have the rodman
hold his rod at A, taking care to hold it vertically, take
the reading of the rod and enter it in the column of
Back Sights opposite station A. In the example it is
5.10. Add the back sight to the elevation of the point
A and we have the elevation of the line of sight through
the instrument or the "height of instrument," as it is
called, and indicated on the notes as H.I. This is
15.10. Enter it in the H.I. column opposite station
A. Next take a sight on B, called a foresight, and
enter the reading in the F.S. column opposite station
B. This in the example is 2.80. Subtract this reading from 15.10, the height of the instrument, and write
the difference, 12.30, in the elevation column opposite
station B. This is the height of B with reference to
the starting-point A. If the elevation of other points
is desired before a change of instrument is made, take
as many foresights as wanted and obtain the elevation
of the points in the same way as just described. Next
change the instrument to some point beyond B, and
take a back sight on station B. Add this reading to
the elevation of B to obtain the height of the instrument in its new position. Enter the sum in the H.I.
column opposite B. In the example the back sight is
3.70, elevation 12.30, H.I. 16.00. Take a foresight
on C and subtract the reading from 16.00, the H.I., and
obtain 13.80, the elevation of C. Remove the instru-

ment to a point beyond C, and obtain the elevation of D in the same way. The points upon which two readings are taken are called *turning-points*. All others except bench-marks are called intermediates. Pegs should be driven upon which to make turning-points. These are frequently called "hubs" in practice. By carefully observing the figure here given, and comparing it with its accompanying notes, the routine of simple levelling can be clearly comprehended by the reader. The column of elevations shows at a glance the comparative height of every point taken with reference to the datum used.

It will now be understood how levelling is simply finding how much higher or lower one point is than another. To insure correct results the instrument should be in good adjustment, rod readings should be taken correctly and entered accurately in the notes, the book work should be correctly carried out, and not the least important matter to be observed is that the instrument should be set about equally distant from both turning-points.

The notes may be proved, first by reviewing carefully all of the additions and subtractions, and second by adding the column of foresights and the column of back sights. Take the difference of these sums, and if it equals the difference of the elevations of the points compared, the work on the book is correct.

LEVEL NOTES TO ACCOMPANY FIG. 11.

Sta.	B S.	H.I	F.S.	Elev.
A........	5.10	15.10	....	10.00
B........	3.70	16.00	2.80	12.30
C........	5.40	19.20	2.20	13.80
D.......	....	....	4.25	14.95

Proof: Difference of elevation between A and $D = 14.95 - 10.00 = 4.95$.

$$\text{Sum of back sights} = 14.20$$
$$\text{Sum of foresights} = 9.25$$

$$\text{Difference of elevation} = 4.95$$

Drainage Topography.

Drainage surveys involve the collection and representation of such facts relating to the surface of the land, its soil and subsoil, as will be of service in determining upon and carrying out a plan for the profitable drainage of the land so examined. It is a branch of work that comes within the special province of the drainage engineer, and it must be done with greater or less thoroughness before he can plan a drainage project with any assurance that it will accomplish the desired work when carried out.

The completeness of a drainage survey must be measured by the time and money that can be devoted to it, and by the thoroughness of the drainage work for which the survey is made. There should be an adaptation of these factors to the result aimed at. The engineer should understand what data and information are required, and then make his plans to obtain them in the most systematic and expeditious manner possible, otherwise he will fritter away his time and energy upon matters that do not pertain to the case in hand. A survey may be required of a field, farm, district, township, or county, yet in all cases the work must be conducted so as to cover the points that will be required within the limits of the allotted time and expense. Every thorough-going engineer takes pride in making his work as complete as possible, even at the expense of using more time and labor than he may be paid for.

The Preliminary Survey.

The preliminary survey or reconnoissance consists in making a personal examination of the ground with reference to its general features or geography, using for this purpose any surveys or maps that can be obtained and information that can be gathered from residents and others who may be acquainted with the land. It should include an examination of the water-courses and ditches, where their source is, and where they discharge. The kind of soil may often be read from the character of the vegetation. The object of this "reviewing" is to determine the practicability of some proposed drainage scheme, or to plan for a more complete survey of the field or tract. If it is simply a field, a few levels may be taken, and the water-shed lines be determined, when the engineer can at once make his plans and proceed with the location work. This is the simplest form of outline survey and is applicable only to fields whose drainage limits and slopes are easily determined or in the case of extended tracts, for the purpose of planning for a more complete survey.

There are several methods of making a more detailed survey, and the value of each will depend upon the nature of the tract and the object sought by the survey. The following are some of the plans which are adapted to this class of work.

With Boundary Line as a Base.

Having the boundary lines of the tract, go to the supposed lowest point and establish a bench from which to level. Run a line of levels on or near the boundary of the tract, taking elevations of the highest and lowest points, important ditches, ponds, etc. It will be best to measure the distances, setting hubs at every 500 or 600 feet and note all points at which levels are taken by the distance they are from the initial point. All elevations should be referred to the initial bench, which for convenience may be recorded as 100. From any station on this boundary, base lines can be run to the interior and any desired point can be located and its elevation taken, or interior cross lines can be run from which the topography can be made up. During this operation the engineer should keep his eyes open to every peculiarity of the land over which he passes. He should keep running notes in his mind and make entries in his book of observations that may be of use to him in making up his topography.

For a survey of this kind, and in fact for any method used in topographical surveys, the rodman should be efficient and expeditious, thus leaving the man at the instrument free to give his entire attention to his proper work without being harassed by a blundering rodman. Care should be taken to use every possible check on the levels so that the engineer may be confident in the end that they are correct.

Water-course as a Base.

A second method especially applicable to district work is to use the main water-course as a base and refer all other lines to it.   In this case the line should be measured, stakes set at the angles and levels taken. Ordinates may be run from this line and in such directions as the judgment of the engineer may dictate. The object is the same in either case, viz., to get the course and slope of the natural depressions, to find the water-shed lines and the area of the drainage-basin.

Method by Central Base Line.

Many tracts of land have such irregular boundaries, with no well-defined outlet stream or other prominent features, that the methods previously given for making a preliminary survey will not be as easy of application as the one to be described.   Run a central base line through the longest dimension of the field or plantation, setting stakes and solid hubs at distances of 400 feet. (See Fig. 12.)   These are to be used as permanent stations in all subsequent work of a preliminary nature, and the line should be described from a compass bearing.   Levels should be taken upon each hub and used as bench-marks.   All of the low and high points may now be sought out by inspection or with the aid of the level and marked by a stake and hub.   Find elevation of these points.   Take a compass bearing and measure the distance from these points to the nearest station on the base line, or if a circle without compass needle is used, set up at the nearest station and turn off the angle from base line to the new points. Each one may now be used as a station in planning and locating the drains immediately surrounding it. The data taken in

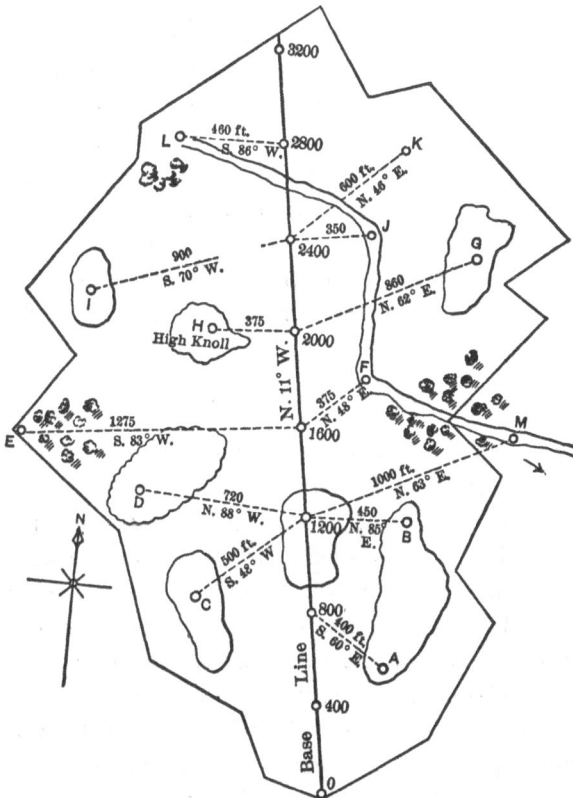

FIG. 12.—Topography from Central Base Line.

this work should be sufficient to plat the drainage system correctly, or if only preliminary levels are desired the field points can be platted. As many points may be located and levelled as may be necessary to plat the

topography of the ground or to locate needed drains.
It is better to keep this work platted up every day, for
while the notes are fresh in the mind, quicker and better work can be done in mapping than if the notes
alone must be relied upon.   While the notes may and
ought to be quite full, the memory enlivens the view
and enables one to add that to the descriptive map
which will be of value.   If there should be no plat of
the tract, the boundary should be run in with the compass if it is desired to have a complete map in every
detail.   All maps should be made carefully to a scale,
the practical value of which will appear when it is desired to make estimates from them for further work.

Record of the Work.

The engineer now wishes to put his work on record
so that he can plan future location surveys intelligently
or represent the capabilities of the tract plainly to
others, which latter consideration is very important.
For this purpose the plat should be transferred from
the field-book and drawn to a convenient scale upon a
sheet of paper and the necessary items entered.   Record the elevations directly upon the map at the location where the levels were taken.   Indicate water-
courses, ponds, trees, etc., by conventional signs.
(Fig. 23.)   Sketch in dividing lines on water-sheds and
indicate surface slopes by arrows.   Designate each
drainage-basin which has a distinctive outlet as a drainage section by the letters A, B, C, etc.

The work thus far done forms a geography of the
tract which shows its natural drainage slopes.   The

elevation numbers show the actual fall from one point to another. The engineer can now compute the area of each drainage section. The topography is plain to every one who may have occasion to examine the map. It fits the ground so that most of the points can be relocated from the map by their relation to natural objects and established features. An approximate estimate of the cost of drainage can be made from this map, though in order to arrive at an accurate estimate, drain lines should all be measured.

Topography by Contour Lines.

Contour lines are drawn upon a map defining points on the surface of the land which have the same elevation. The vertical distances between these lines may be taken at 6 inches, 1 foot, or 2 feet, or any other desirable distance, in which case the number of the contour lines show to the eye at once the elevation of the ground over which the line passes. The line of greatest slope of the land will, of course, be directly across the contour lines.

For the purpose of illustration, suppose it is desired to make a survey and map of a farm from which a plan for its drainage is to be made and in time executed. The earlier in the work such a map can be made, the greater will be its value. Especially will such a map be of service if the projected work is to be done at different times as facilities or means may permit.

The Survey.—Begin at one corner of the farm or field whose adjacent sides are straight lines and use

these two sides as bases from which to work. Have
stakes prepared which should be about 16 inches long.
Common lath are good for this purpose. Begin at the
corner and measure off a base, setting a stake at each
station of 100 feet. Letter the stake at the corner *A*
and the others *B*, *C*, *D*, etc., in order. This is the base
line. Begin at the point *A* and measure from that
point along the adjacent side of the field, and number
these stakes 1, 2, 3, etc., until the limit of the field is
reached. The last stake should record the length of
the line in feet.

Set a flag 100 feet from the last stake at a right angle
with the line run, so that a line can be run parallel with
the first. Begin at the stake *B* on the base line and
measure a line and set stakes parallel to the first line.
Proceed in the same way across the entire farm until
it is entirely checked into squares of 100 feet. In laying off these lines, they should be kept straight by
means of flags, which are set ahead of the work, and in
case a prominent feature, such as a centre of a pond or
a stream is crossed, an intermediate stake should be set
and properly numbered. These lines can now be described as *A*, *B*, *C*, etc., lines, and any point on that
line by the number of the stake on it.

The next work is to " book " the farm or field, as the
case may be. For convenience it will be best to select
the lowest point on the farm as a datum if it is apparent to the eye. If not establish a " bench-mark "
and assume a datum plane at the initial point or *A* of
the base line. Take each line in order and take a
level at each stake, recording the elevation of the surface under its proper head. The headings of the level

pages would read "Levels on Line A," "Levels on Line B," etc.

The stations at which levels are taken should be numbered as they are on the stakes. Additional natural features of the surface should be noted on the book in connection with the elevation of the stations. When the whole farm has thus been gone over, the level book will show the elevation of the ground at the position of every stake that has been set, which forms the data from which a map is to be made.

Practical Hints.—Before leaving the description of the field work several practical hints will not come amiss. The staking may be done by two active young men, one at each end of a steel band chain, if it is preferred. The head chainman carries the stakes in a hand basket and sticks one at the end of the chain, the rear chainman lining him in by a flag-pole, which has been previously set at the proper place. The rear chainman numbers the stake properly and drives it with a hatchet. The stakes at the boundary of the farm should be permanent ones and remain in position so that the points in the interior can be produced at any time desired.

When the levelling is being done, two lines may be taken at one setting, and upon completing the first two lines the next two may be taken from the upper end back toward the base line, provided care be taken to keep the notes in order so that each station shall have its proper elevation recorded against it.

"Bench-marks" should be established at convenient places for future reference. "Turning-points" should be taken on pegs, but other levels may be taken from the ground surface.

The work should be done at a time when growing crops will not interfere with running the lines.

It would appear upon first considering work like this, especially to one who has not tried it, that obtaining the data before described would involve considerable time and labor. It may be said, however, that 160 acres of ordinary prairie farming land may be surveyed as above described—two chainmen with stakes, followed by levelman and rodman—in two days.

The Map.—All necessary information is now in tabular form on the field-book. Make a map of the farm by first adopting a scale and platting the boundary lines according to notes and measurements taken in the field and recorded in the book. One half inch to 100 feet is a good scale to use for a farm of 160 acres. Reproduce the lines laid off in the field so that the plat will correctly represent the field on the scale adopted and used. (Fig. 13.)

Now write the elevations as found in the field-book at the intersections of the lines on the plat, which intersections represent the position of the stakes which were set in the field.

The plat now shows the comparative elevations of these points over the entire farm, and also such other features as may have been noted in making the survey. The contour lines may now be drawn from the elevations which are shown. The vertical distance between them may be any unit it may be convenient to adopt. In the example here given, the distance is one tenth of a foot. They are numbered in order of elevation. The plat now shows at a glance the degree and direction of slope of any part of the tract. The elevation figures need not be retained after the contours are sketched in and numbered.

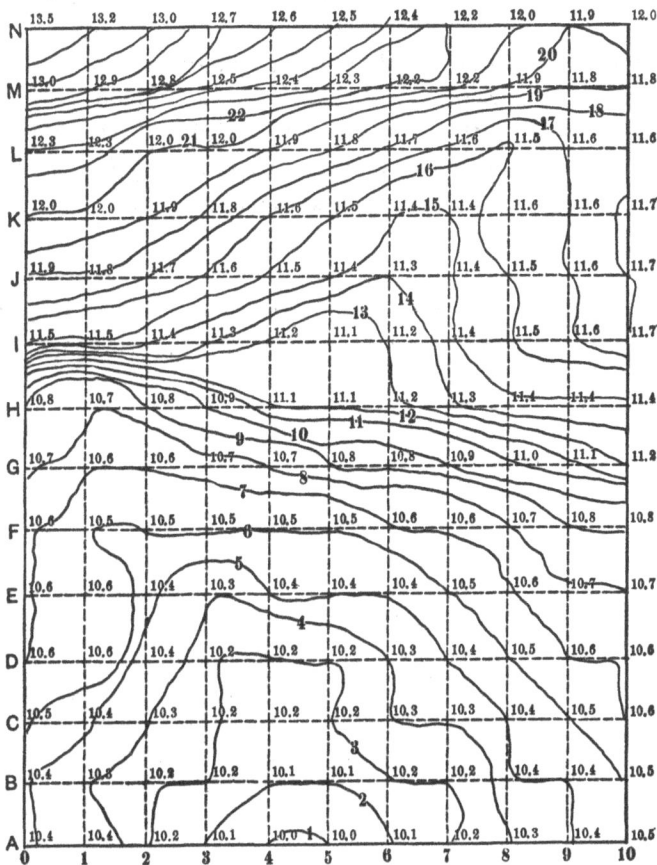

FIG. 13.—Topography by Contours. Squares 100 feet, Level Datum 10 feet. Interval between Contour Lines $\frac{1}{10}$ foot.

This method of representing topography is so complete that a plan for drainage may be accurately laid upon the contour map. The contours show points of equal elevation over the entire field in such a way that the surface slope is shown to the eye at a glance. However, for ordinary drainage work the topography of a field may be found in sufficient detail by methods previously described which require less labor. A knowledge of the elevation and location of the high and low points in a field or district is usually sufficient for the effective planning of a drainage system.

CHAPTER V.

LAYING OUT DRAINS IN THE FIELD.

Staking Out Drains.

ONE or more lines of drains, or an entire system having been determined upon, the next work is to stake out
the lines and prepare them for the construction of the
drains. Stakes should be prepared beforehand. A
good material for stakes is what is known as fence lath.
They are 4 feet long, 1½ inches wide, and ⅝ inch thick.
They may be cut in three pieces, making them 16
inches long, which is a suitable length for land which
is reasonably free from grass and heavy weeds, but
ordinarily they should be 2 feet long. These are
called guides, and serve to carry the necessary figures
and to show the location of the grade stakes. An
equal number of grade stakes should be made to accompany them. They may be of the same material,
but only 1 foot long. Prepare as many sets of these
as there are stations of 100 feet to measure off, with
some extras for intermediates. Where the work of
making the ditches is to be done without much delay,
common plastering lath, which are more easily carried,
may be used for guides.

Begin with a main drain, first flagging out the course
so that the stakes may be lined in straight. Set the

first stake near the outlet of the main and about 14 inches to the right of the centre of the proposed drain. Drive the grade stake to the surface of the ground, and the guide stake about 4 inches beyond it, as shown in Fig. 14. Let the fore chainman hold the forward handle of the chain and a guide stake in a vertical posi-

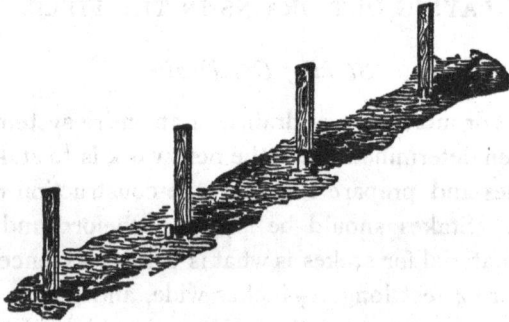

FIG. 14.—Guide-stakes and Hubs.

tion in the same hand, and let the rear chainman, with the handle of the chain at the grade stake and his eye directly over it, line him in with the flag which marks the position of the line to be staked out. The fore chainman sticks the stake where directed and drops a grade stake by it. He then pulls ahead another length of the chain and is again put into line. The rear chainman drives the stakes and marks them with a large lead-pencil in order, calling the first stake o and the next 1, etc., for all full stations from outlet to the upper end of the drain.

Where curves are made, intermediate stakes should be set in such a way that they can be followed and used in digging the ditch, and should be marked so as to indicate the number of feet from the outlet up to each stake. As for example, between stations 5 and 6 curve stakes are set 20 feet apart. They would be marked 5.20, 5.40, 5.60, etc., and indicate $5\frac{2}{10}$ stations, or 520 feet, etc.

Another thing to be noted at the time of staking the main is where the sub-mains and branches enter. If it is desired that a branch line should join the main between stations, a stake should be set as an intermediate with the character o upon it and also the name of the drain that enters.

The same plan of staking out lines of all kinds should be followed. Begin at the junction stake, and set and number the stakes from the o or junction point upward until the upper end is reached. The stake at the upper end of each drain should have its respective name upon it in addition to the station number, so that in looking over the system the drains can be followed from either end by schedule or map.

Designating Drains.

Some system of designating drains is needed where there are many of them in a system, in order that the notes may be kept without confusion and also correspond with the schedule and plat which should be made after the work is laid out.

Mains may be designated as main A, main B, etc., in the order of their size or importance. Branches of each main may be numbered in order, from the outlet of the main up, as No. 1, No. 2, etc. If one of these branches is a sub-main, that is, one receiving laterals, the laterals may be designated as a, b, c, of that sub-

main.   All numbering and lettering of drains is done
from the outlet toward the upper ends.   If this system
is not followed, some other equally clear one should
be used, so that the engineer may avoid the possibility
of applying the notes of one drain to some other to
which they do not belong.   When the whole arrangement is made clear, the contractor may take the work,
and without any further explanation than that which
appears in the field and the schedules, he may follow
out the work in all its details.

Levelling Drains and Keeping the Notes.

The method of taking the levels after the drains are
staked out is the same as that previously described.
The rodman should hold the rod in a perpendicular
position upon each grade stake in order beginning, at
o stake, and after the " all-right " signal from the
levelman, he should at once pass to the next stake
and " rod up " for another reading.   He should call
off the station numbers of all intermediate stations,
branches, etc., as he approaches them that the levelman may enter them correctly upon his book.   It
should be remembered that in this kind of work each
level should be taken with the same care.   Every stake
must be worked from in digging the ditch with equal
care.   Sights of 400 feet in length should be regarded
as about the limit, though longer ones can be taken
with reasonable accuracy.   The levelman should see
that his instrument is all right at the time he reads the
rod, and he should keep his mind closely upon the work
or he will take incorrect readings or enter them incor-
rectly upon his book.

The field-book should have two more columns ruled in addition to those described for simple levelling, one for the elevation of grade line or simply "grade line," ($G.L.$) and one for the depth of cut at each station marked "Cut." These latter columns are for computations to be made after the field work has been completed. The accompanying specimen page of a field-book gives an example of entries made in the field and the subsequent computations for a short drain.

A sketch plat of the lines should be made on the right-hand page. This is very convenient and may with proper care be made quite accurate by means of measurements taken to certain landmarks or permanent objects and by sketching in the angles and curves. All of this can be done at the time of levelling without retarding the work to any extent. Be particular to note where branch lines of whatever kind enter other drains.

Notes for Platting.

These should be taken at the same time the levelling is done and recorded on the right-hand page of the notes if it is expected that a plat better than the one which can be made from the sketch will be required. *If done with a compass*, locate all outlets of drains with reference to some corner of the farm most prominent or convenient by means of a measurement and bearing from it. Take the bearings of the straight lines of the ditch, and if a long curve is made, of several parts of the curve so that it may be represented with reasonable accuracy. Note where fence lines are crossed, and make measurements from convenient stations to bound-

LEVELLING FOR DRAINS—FORM FOR FIELD-BOOK,
MAIN A.

(Left-hand Page.)   *(Right-hand Page.)*

Sta.	B.S.	H.I	F.S.	El.	G.L.	Cut	Remarks
0.....	6.45	106.45	......	100.00	97.25	2.75	Grade .25 to 100 ft.
Outlet.			9.20	97.25	......	......	Bottom outlet ditch.
1.....			6.32	100.13	97.50	2.63	
2.....			5.92	100.53	97.75	2.78	
3.....			5.61	100.84	98.00	2.84	
4.....			5.12	101.33	98.25	3.08	* Change grade .20 to 100 ft.
4.50..	7.10	108.94	4.61	101.84	98.35	3.49	Branch No. 1 enters.
5.....			6.72	102.22	98.45	3.77	
6.....			6.11	102.83	98.65	4.18	
7.....			6.72	102.22	98.85	3.37	
8.....			7.21	101.73	99.05	2.68	Pond.
9.....			7.11	101.83	99.25	2.58	
10....			7.40	101.54	99.45	2.09	
11....			7.10	101.84	99.65	2.19	Ends 25 ft. from S. line of field.

Bearings.

0– 3   S., 3° W.
3– 7   S., 25° W.
7–11   S., 5° E.

Note.

Outlet of drain on line of farm is 350 ft. W. of N.E. Cor.

* The profile of the above notes is shown in Fig. 15.

ary lines of the field or farm in order to check the work in drawing the plat or map. If a level with an attached compass is used, these notes can be taken with little additional effort and time, as use of the compass can be easily acquired. The compass gives all angles referred to the magnetic meridian for that place, which fact makes the work more expeditious than any other known method. *If done with an instrument having only a graduated circle*, some line of the field or farm must be taken as a base, and the angle which the first straight line of the drain makes with it be measured. Then the angles which the several tangents of the drain make one with the other should be measured, being careful to record whether these angles are right or left. The same methods of checking by measurements as the chaining is done should be observed as those described for compass work.

A Common Datum.

Every drain in the system should have distinct notes by itself similar to the example given above for illustration. All elevations should be referred to the same datum, so that the difference of any two elevation numbers anywhere in the system will show the actual difference of elevation of the land. In other words, the levels should all be connected. This can readily be done by observing this simple rule. When levelling is begun on any new line or branch *take the first back sight on some point whose elevation is recorded and add it to the elevation of that point for a new height of instrument.*

The Profile.

A profile represents a vertical section of the line of survey. The level notes are platted on profile paper on which the vertical and horizontal scales are different to render irregularities of surface more distinct through exaggeration. The method of profiling the level notes is shown in Fig. 15. In this diagram, one space, hori-

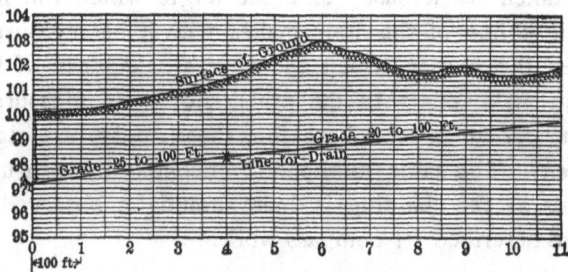

FIG. 15.—Profile of Main A.   (See Notes.)

zontal scale, equals 50 feet, and one small space, vertical scale, equals .2 foot of elevation, which are convenient scales in making profiles for drainage work. A profile represents the surface line correctly and should be employed in fixing upon a grade line when the work to be done is of any considerable magnitude.

To determine a grade, draw a thread over the profile, adjusting it to certain points of maximum and minimum grade, until a desirable grade line is represented by the thread. Transfer the elevation of points where changes of grade occur to the corresponding points on the notes, and compute the cuts for each station. The profile represents the approximate relation of surface to grade line, but does not indicate the cuts at each station sufficiently close for use in detail construction. Every profile should have a name or heading by means of which it can be connected with the notes from which it has been made, or with the surveyed line on the ground which it represents in section. Profile paper used for this work is sold by houses dealing in engineers' supplies.

Compass Surveying for Drainage Work.

The data for platting farm and field lines, locating preliminary level points for topographical work, drain lines, etc., can be most rapidly obtained by use of the compass. A level equipped with a small compass suitable for field work is convenient and very serviceable. The needle indicates the magnetic meridian, an approximately north and south line. The true meridian is a true north and south line, which if produced would pass through the poles of the earth.

The compass circle (Fig. 16) is divided into degrees and fractions of a degree. The letter E., denoting east, is at the left hand, and W., denoting west, at the right hand of the box, which is contrary to the position of these letters in the small pocket compasses. This arrangement is necessary because in using the field compass the box is turned so that the sights point in the direction of the line whose azimuth is to be obtained. The north end of the needle is read, which gives direct the azimuth of the line or the angle which it makes with the magnetic meridian.

The *bearing* of a line is the angle which it makes with the direction of the magnetic needle. The length of a line with its bearing is termed its course. To take the bearings of a line, set the compass directly over a point in it at one extremity if possible, though this is not essential. Bring the compass to a level position. Have a flag or rod set on another point of

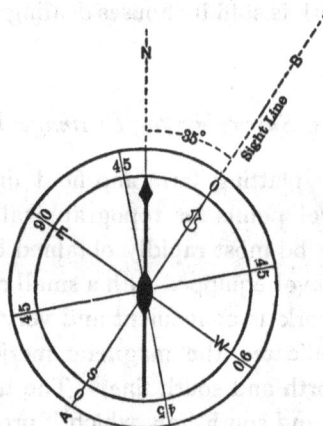

FIG. 16.—Taking Compass Bearings.

the line. Direct the sights upon this rod as near the bottom as possible. Always keep the north end of the compass ahead. It is distinguished from the south end by some conspicuous mark on the face. Sight accurately to the flag and read the north end of the needle. To do this, note first whether the N. or S. point of the compass is nearest the north end of the needle, second, the number of degrees to which it points, and third, the letter E. or W. nearest the north end of the needle. Always read and record bearings in this order.

In the figure the line is AB along which the sights point. The needle points constantly to the meridian, hence in turning the sights to the line AB the angle NB is turned off, or from 0° to 35°, and the needle reads north 35° east, hence the bearing of the line is N. 35° E.

To test the accuracy of the bearing set up the instrument at the opposite end of the line and take a back sight upon the first point. If the reading agrees with the first, but with opposite letters, the bearing first taken was correct. The declination of the needle is the angle which the magnetic meridian and the true meridian make with each other. There is always a declination to take into account except on or near a certain line passing across the country called the line of no variation. The declination of the needle is constantly changing. It is desirable to record lines with their true bearings, or as nearly so as practicable, though this feature of the work does not possess the importance which is attached to it where surveys are made for the definition and determination of land lines. The local declination can be determined by setting up the compass upon an old land line whose bearing is known if such can be found, or in the absence of such a line a bearing may be taken upon the pole star and declination noted. This will be only approximate, as the star is $1\frac{1}{2}$ degrees from the pole, revolving about it, and being on the true meridian only twice in twenty-four hours.

Another method of determining an approximately true meridian is by equal shadows of the sun.

On the south side of a level surface, as at S in Fig.

17, place an upright staff not less than 10 feet long. Two or three hours before noon mark the extremity A of its shadow.   Describe an arc of a circle with S the foot of the staff for centre and SA the distance to the extremity of the shadow for radius.   About as many

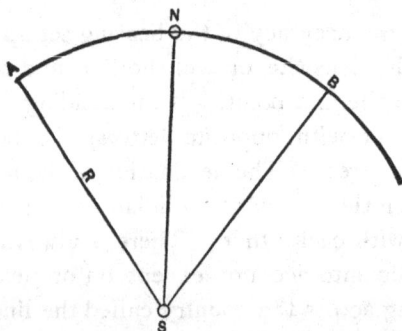

FIG. 17.—Obtaining Meridian by Equal Shadows of the Sun.

hours after noon as it had been before noon when the first mark was made, watch for the moment when the end of the shadow touches the arc at another point B. Bisect the arc AB at N.   Draw SN and it will be the true north and south line.   Set up the compass at S, sight on N or SN produced and read the needle. The reading will be the declination of the needle at that place.   It is more important, however, to record on the notes the declination used than it is to go into the niceties of obtaining and using an absolutely correct declination angle for line work of the character herein described.

If the compass has a variation plate, set off the declination assumed or determined and record all bearings as read.   If there is no such provision for mechanically correcting the azimuth, make corrections on the notes according to the following rule.

To Reduce Magnetic Bearings to True Bearings.

When the variation is east, as in Western and Southern States, for bearings N. and W. or S. and E. subtract declination from magnetic bearing. For bearings N. and E. or S. and W. add declination to magnetic bearing.

When the variation is west, as in the Northeastern States, for bearings N. and W. or S. and E. add declination.

For bearings N. and E. and S. and W. subtract declination.

Now that wire fences and other improvements made of iron and steel are in such common use, care must be taken to keep the compass distant from all iron and steel, which of course will deflect the needle and destroy the reliability of its readings. Should it be necessary to obtain the bearing of wire fence line, an offset of 30 feet may be made and the bearing of the parallel line be read.

Keeping Compass Notes.

The running form of keeping notes is simple and in common use. For example, in recording the notes of drains, the following running notes may be written on the right-hand page of the level-book.

DRAIN NO. 2.

Sta. 0 to 6................N. 10° 30' E.
" 6 to 8............ ......N. 4° 00' E.
" 8 to 15................N. 32° 00' W.
" 15 to 22 (end)...........N. 15° 20' W.

If certain points for mapping or topography are to be located from some station, the following tabulated form may be employed:

OBSERVATIONS AT PRELIMINARY STATION NO 6

Sta.	Bearings	Distance Feet	Remarks.
Sta No. 6 to			
A	N 42° 30′ E	800	Low swale
C	S. 70° 20′ W	1000	Walnut-tree
D.............	S. 60° 00′ W	600	Upper end ditch to horse pond.
E	N. 20° 30′ W	800	Highest point in west field

The same form may be used to record a continuous and connected line like the boundary of a farm or field

Back sights should be taken at each station to ascertain if there are any disturbing influences which cause the needle to read differently at the two ends of the line. If a discrepancy in the two readings is found, some point on the same line intermediate between the two should be used to determine which of the bearings is correct.

The instructions thus far given are sufficient to suggest to the beginner the method of doing simple line work with the compass. With a little practice in the field, without which no one can understand the subject fully, a novice may soon acquire all needed proficiency.

CHAPTER VI.

FIXING THE GRADE OF DRAINS.

A grade line as determined by a survey is the line along which the water of the drain, when constructed, is to flow.

Fall is the common term for slope of surface when applied to land, and for total head when applied to drains.

The available fall is the fall that can be given to a drain in a prescribed distance, as distinguished from the fall of the land through which the drain extends.

The grade of a drain is its rate of fall and is expressed in inches, or in decimals of a foot, per 100 feet. When expressed in decimals of a foot per 100 the grade is said to be so much per cent.

Determining the grades upon which the drains should be laid requires much skill and knowledge of practical details of construction, together with the understanding of the requirements of the soil, capacity, cost, and efficiency of various kinds and sizes of drains, some of which subjects will be discussed in subsequent chapters of this work. The minimum grade that may be successfully used for the tile drains is a matter of great moment where level lands are treated, and will depend much upon the accuracy with which the drains will be constructed. The topography of the surface

85

places limitation upon grades which cannot always be changed by artificial work.   While grades of 2, 3, or 4 inches per 100 feet may be desired as a minimum, they cannot be obtained in very many tracts of land which may be successfully drained.   Grades as low as $\frac{1}{2}$ inch per 100 feet are in successful operation, giving good results on thousands of acres of land.   Mains laid for a distance on a level are sometimes used with success, the flow through such mains depending upon the head given by the free water in loose soils, and by lateral drains having a grade greater than that of the main. The lack of fall must be offset by increased size of drains, and by the greatest degree of accuracy in their construction.

A uniform grade is the simplest.   Having decided upon the depth of drain at the outlet and also at the upper end, as for instance 3 feet, subtract this from the elevation of each of these points and obtain the elevation of the grade line at the outlet and upper end respectively.   The difference of the elevation is the available fall, which, divided by the number of stations, gives the fall per station.

Starting with the elevation of the grade line at the outlet, add the grade per station to this elevation to obtain the elevation of the grade line for the next succeeding one, and so continue to add the increments. For intermediate stations use a proportional part of the grade per station.

The cut or depth of drain is found by taking the difference between grade and surface elevations.   The last column of figures under the head "Cut" is the object sought, and is that for which all other work so far has been done.   The cut measured from the top of the grade pegs and connected by a line will give a true and uniform grade from beginning to end.

A change of grade is frequently necessary where the land has any considerable slope, otherwise the drains will be too deep or too shallow in places for economy in digging or for efficiency of operation. This is effected by dividing the lines into the necessary divisions, each division having a grade of its own. The station where the change of grade is made should be noted on the book.   In the example of the notes given the grade is .25 per 100 as far as station 4, and then changes to .20 per 100.   Where a cut is to be made through a ridge to reach a flat which it is desired to drain, determine the least depth of drain that should be used at the upper end, take a safe grade, say .10 per 100, or .20 per 100, and run down by subtracting the grade from the elevation of each station in order, until the ridge is passed and the desired depth is obtained, then change to a heavier grade.   It is the ordinary method of grading a drain reversed.

A large number of examples might be given, but the above are sufficient to show the beginner the general plan of work.   A few examples of his own worked out will soon give him an insight into the practical details of grades.

When a sub-main or lateral enters another drain it is best to have an outfall from the branch line into the main.   This is commonly called a " drop " and should be proportionate to the size of the tile used on both lines.   For example, branches into a 6-inch main should drop .20, into an 8-inch .30, 10-inch .40, 12-

inch .50.   To compute this, add the drop to the elevation of grade line at the point of junction, which will give elevation of starting-point of branch.   Example: At station 4.50 of notes, Branch No. 1 is to have a drop of .20.   The grade line 98.35 + .20 = 98.55 = elevation of grade line at outlet branch.   This

FIG. 18.—Angle and Drop for Tile Drain.

should be transferred to the notes of Branch No. 1 and used as the initial point for computing the grade of that line.   If short bends or curves are necessary a little additional grade should be allowed, provided it can be had.

The depth at which it is desired to lay the drains will often have much to do with the determination of the grades.   No inflexible rule can be given, but uniformity should as much as possible be secured.   If the flow in drains is alternately slow and rapid and then stands still, the tile being full at one place and half full at others, the efficiency of the drain is only part of what it might be were the grades carefully arranged and the size of the tile proportioned to them.   A careful survey and corresponding adjusted grade will often add one half to the efficiency of a drain, as compared with one which is carelessly made.

For convenience in reducing the "Cut" column of the notes to feet, inches, and fractions of an inch, which will usually be demanded by workmen in digging a ditch, a table is here appended. In all engineering computations it is desirable to use the decimal scale, but the engineer will soon learn the equivalents of decimals of a foot in inches and fractions so that he can write them without referring to the table. Reductions given to the nearest ⅛ inch are sufficiently close for practice.

TABLE 1.

DECIMALS OF A FOOT REDUCED TO INCHES.

Foot.	Ins.	Foot.	Ins.	Foot.	Ins.	Foot.	Ins.	Foot.	Ins.
.0104	⅛	.2188	2⅝	.4271	5⅛	.6354	7⅝	.8438	10⅛
.0208	¼	.2292	2¾	.4375	5¼	.6458	7¾	.8542	10¼
.0313	⅜	.2396	2⅞	.4479	5⅜	.6563	7⅞	.8646	10⅜
.0417	½	.2500	3	.4583	5½	.6667	8	.8750	10½
.0521	⅝	.2604	3⅛	.4688	5⅝	.6771	8⅛	.8854	10⅝
.0625	¾	.2708	3¼	.4792	5¾	.6875	8¼	.8958	10¾
.0729	⅞	.2813	3⅜	.4896	5⅞	.6979	8⅜	.9063	10⅞
.0833	1	.2917	3½	.5000	6	.7083	8½	.9167	11
.0938	1⅛	.3021	3⅝	.5104	6⅛	.7188	8⅝	.9271	11⅛
.1042	1¼	.3125	3¾	.5208	6¼	.7292	8¾	.9375	11¼
.1146	1⅜	.3229	3⅞	.5313	6⅜	.7396	8⅞	.9479	11⅜
.1250	1½	.3333	4	.5417	6½	.7500	9	.9583	11½
.1354	1⅝	.3438	4⅛	.5521	6⅝	.7604	9⅛	.9688	11⅝
.1458	1¾	.3542	4¼	.5625	6¾	.7708	9¼	.9792	11¾
.1563	1⅞	.3646	4⅜	.5729	6⅞	.7813	9⅜	.9896	11⅞
.1667	2	.3750	4½	.5833	7	.7917	9½	1.00	12
.1771	2⅛	.3854	4⅝	.5938	7⅛	.8021	9⅝		
.1875	2¼	.3958	4¾	.6042	7¼	.8125	9¾		
.1979	2⅜	.4063	4⅞	.6146	7⅜	.8229	9⅞		
.2083	2½	.4167	5	.6250	7½	.8333	10		

Depth of Drains.

The depth which drains should be laid is a matter which has received a great deal of attention since the time that underdrainage began to be practised. Advocates of deep and shallow drains have very earnestly

argued their favorite theories.   It is one of those cases
in which theories do not always work out in practice,
the factor which prevents this being the variations in
the characteristics of the soil which is treated.   In order that any one theory may prove correct, it must
be assumed that a soil of a certain kind under certain
conditions is to be operated upon.   This kind of soil,
however, is not always present, and the theory cannot
apply in full.

In speaking of depth of drainage, 4 feet is called deep
drainage, 3 feet medium, and 2 to 2½ shallow drainage.   If drains are laid deep the soil must be susceptible to the ready percolation of water, and by this
process be converted into a soil of greater or less value

FIG. 19.—Effect of Depth of Drains on Open Soils.

to plants.   In Fig. 19 the difference in depth is shown
to the eye with its attendant advantage.   Another advantage is that the soil has a greater reservoir capacity
for water which is valuable in times of excessive rainfall, and still another, the drains may be varied in distance apart upon the principle illustrated in Fig. 9.
Now, this is all true for deep, permeable, rich soils,
and with such there is no doubt as to the value of general 4-foot drainage.

On the other hand, many subsoils at a depth of 4
feet have no fertility in them.   Though plant roots

often penetrate them seeking moisture, they are quite retentive, so that drainage-water passes through them slowly. In such cases drains of less depth than 4 feet are of greater value for agricultural purposes. When the statement is made that drains should never be laid as shallow as 2 feet, it is confronted by the fact that in many localities where the soil is exceedingly retentive, and the subsoil more so, deep drains have little immediate effect. Not that they are devoid of value, or will not in time prove beneficial to the soil, but their value will not be commensurate with their cost. The more retentive the soil, the steeper will be the line of saturation, and the less will be the breadth of land which will be acted upon by each drain.

It may be said that for farm lands, lateral drains should be about 3 feet deep, unless the compact and retentive soil indicates that less depth should be used. When it is attempted to follow any general depth the necessity of obtaining suitable grades for the drains will often make some parts of the drain deeper or shallower than desired. A nice and economical adjustment of the depths of the several drains of a system can be learned only by practical work. A practical knowledge of the field, coupled with the facts on the field-book, form the key to the dormant resources of the soil.

Frequency of Drains.

This is also a question upon which there is a wide difference of opinion and a consequent difference of practice. The science and art of land improvement

are peculiar in one respect, and that is this: No rule or
plan applicable to one locality will strictly apply to another. A design for a park in one city will not be
suitable for another, owing to varying natural features,
as well as requirements which must be met. A soil
in one locality will drain as readily and perfectly with
drains 150 feet apart as others will with drains 40 feet
apart, and upon this fact depends the distance apart
that drains should be placed. It would be a waste of
labor and material to place drains 40 feet apart in
some of our soils, while, on the other hand, to place
drains at intervals of 150 feet in some soils would
come far short of accomplishing thorough drainage.

The cost of the work has much to do with the dis-

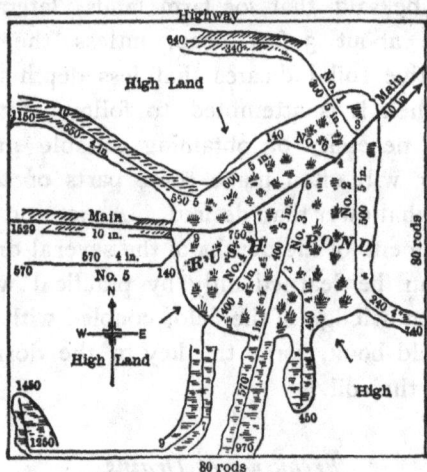

FIG. 20.—Drainage of a 40-acre Field containing
Pond with High Land surrounding it.

tance apart at which drains are usually placed, and
thorough drainage is often sacrificed to this. As ob-

served in another chapter, the object of thorough drainage is to bring all the soil under the influence of drains either natural or artificial. There are soils where drains 200 feet apart give good drainage for farm crops. There are others where 33 feet is none to near to lay the drains. These conditions vary so widely that one familiar with one kind of soil is inclined to disbelieve a statement regarding the other. Here is where the experience and close observation of the engineer come into use and should be worth many dollars to the landowner. It cannot be urged too strongly upon the engineer who is entering upon this class of work, to familiarize himself with that very interesting subject, the behavior of soils under different methods of treatment and also acquaint himself with the physical differences of soils with reference to drainage.

This is a field for the exercise of close observation upon the ground. As a hint along this line, it may be said that the vegetation is a good index to the natural character and condition of the soil. Certain plants grow luxuriantly upon some kinds of soil and not on others. Learn what these are and keep them in mind when reviewing the land. There may be some open ditches of greater or less magnitude which affect the adjoining land more or less widely as to its drainage. This effect may be known by the appearance of the land and vegetation which is found upon it. If it is the spring of the year, dig a post-hole and note how rapidly it fills with water. These observations may suggest the means by which may be gained that knowledge of the soil which is necessary to an intelligent location of drains as to depth and distance apart.

Preliminary Estimate.

The following will be useful in making preliminary estimates of the number of feet of drains which will be required per acre when laid in parallel lines at the distances apart indicated:

20 feet apart.	2178 feet
25 " "	1742 "
30 " "	1452 "
40 " "	1089 "
50 " "	872 "
100 " "	436 "
150 " "	291 "
200 " "	218 "

The number of feet of drains per acre as shown above does not include any intercepting main which may be necessary to make the work complete. For instance, should it be necessary to locate a main through the centre of a field, its length must be divided by the number of acres in the field, and the result added to the number which is found in the table above, opposite the number in the column indicating the distance apart which it is proposed to lay the drains.

CHAPTER VII.

MAPS AND RECORDS.

THE drains having been staked out, the grades and cuts figured, and the size and number of the tile fixed upon, a map of the drains should be made which will show their position, length, fall, size of tile, and the physical features of the land through which they pass. The details of farm work are usually executed upon the ground so that complete data is not secured until the work is finished.

A sketch map can be readily made from the notes which were taken in the field, and will show quite approximately the position of the lines, and should be made and used as a working map in the distribution of tile and in digging the ditches. A copy of the depth figures for each line, with the working map, constitutes the information which will be necessary for any man or set of men to construct the work as laid out.

The finished map should be made from measurements and angles which were taken in the field for that purpose, and should be drawn to a scale and sufficiently embellished to present a creditable appearance. The young engineer who has had no previous training or practice in this work should not fail to take up this branch and study to make his maps and records creditable and accurate.

Drafting Instruments.

While there are other instruments which are sometimes desirable, the following are all that are essential for ordinary work: A right-line pen, a scale divided to tenths of an inch, a drawing-board 24 inches square, a bottle of liquid India ink, a protractor for platting angles, a T square with 24-inch blade, a bottle of carmine ink, a few thumb-tacks for fastening paper to the board and some good steel pens.

A serviceable paper for making working plats is what is known as bond paper. The size of sheet most convenient is 18 inches by 24 inches. The merit of this paper is that it is flexible, does not crack when folded and carried in the pocket, is partially transparent, so that it can be used in making tracings, and also constitutes a fair negative from which blue prints can be made. Vellum or tracing-cloth is particularly adapted to use in making duplicate copies where fine blue prints are desired.

For finished maps, Whatman's hot-pressed, unmounted paper should be used; sheets 17 × 22 are convenient in size.

Platting Compass Notes and Angles.

The platting of a survey made with the compass consists in drawing on paper the lines and angles which have been measured on the ground. The lines should be drawn to scale and the angles measured with a protractor. A protractor is a scale in the form of a semicircle of brass or celluloid divided into 180 parts or degrees and numbered in both directions. The straight edge has a mark in the middle opposite the 90° mark on the circumference.

To lay off any angle at any point place the straight edge of the protractor on the line with the mark at the point; with the point of a sharp pencil make a mark on the paper at the required number of degrees and draw a line from the mark to the given point.

To plat compass bearings draw a meridian line in light pencil through the initial station or starting point, place the protractor upon the line and point as directed in the preceding paragraph, and lay off the angle called for by the notes. Set off by scale the distance on this course to the next point. Draw a meridian line through the point thus established and in the same manner plat the next and following courses. Should a field have been surveyed, the last course

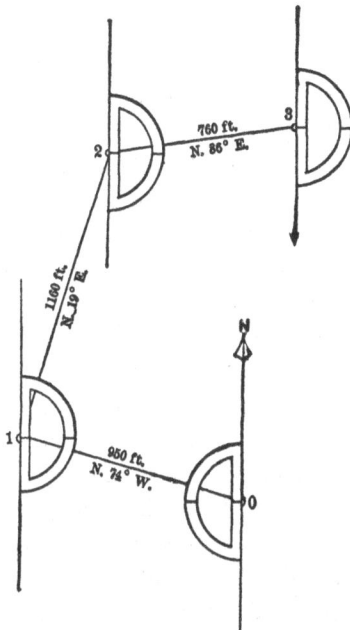

FIG. 21.—Platting Compass Bearings.

should end at the starting-point and the plat should come together or close. If it does not close, it shows that some error has been made either in the field or in platting the work. The method of using the protractor

in platting is shown in Fig. 21. The semicircle part
of the protractor should be placed in the direction of
the course to be marked and the angle read from the
north end of the protractor for all bearings beginning
with *N* and from the south end for bearings beginning
with *S*.

Making the Map.

Determine first how large it is desired to have the
map and the scale that can be used. The measurements in the field have been taken and recorded in
stations of 100 feet, so a convenient scale for representing the lines will be a certain number of hundred feet
to one inch. The more detail work it is desired to
represent on the map the larger should be the scale.
200 to 300 feet to 1 inch are good scales for farms
of moderate size, while for large tracts 500 to 1000
feet must be used in order to keep the size of the
map within convenient limits. Lay off the boundary
of the farm or tract according to the proposed scale.
Locate the position of the outlets of the drains and
work upward in laying off the lines in the same manner and order as the survey was made. Number the stations where angles occur, where branch drains enter,
and also the number of the station at the upper end of
each drain line. All angles should be laid off with the
protractor, and the intersection of the drain lines with
field or fence lines, and in large tracts with land lines,
should be shown. All of this outline work should be
done in light pencil lines, and when completed re-
drawn in ink. In working maps the drain lines are
usually drawn in red, all others in black. For a finely

finished map, it is better to indicate the drains by broken black lines.

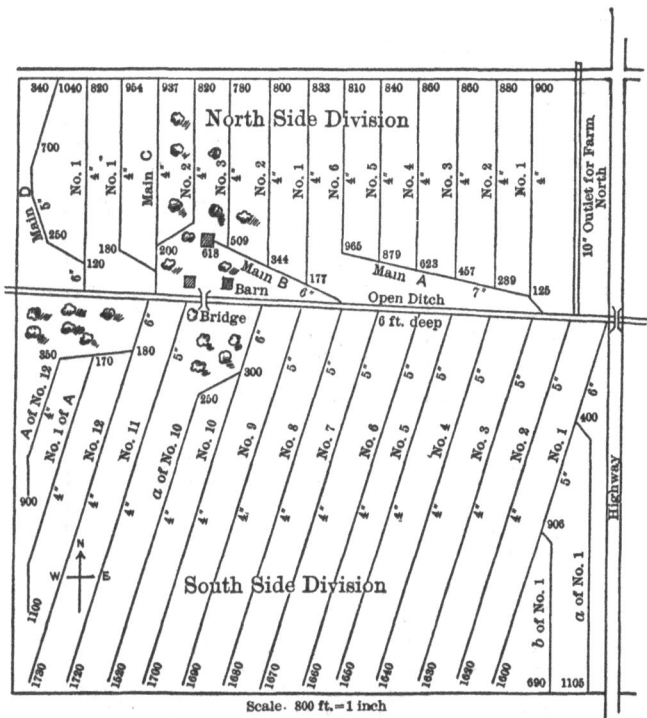

FIG. 22.—Drainage Plan for a Farm of 160 Acres of Level Land.

Conventional Topographical Signs.—These are representations of some of the leading features of the land by arbitrary signs which resemble the objects as we look down upon them. A sufficient number of these should be used to indicate to the eye at a glance what the character of the surface is. This representation should be aided by words of description where full details are desired to be shown.   Peculiarities of the soil may be indicated in this way and thus useful facts be recorded.   A few of the more common conventional topographical signs have already been shown in Fig. 23.

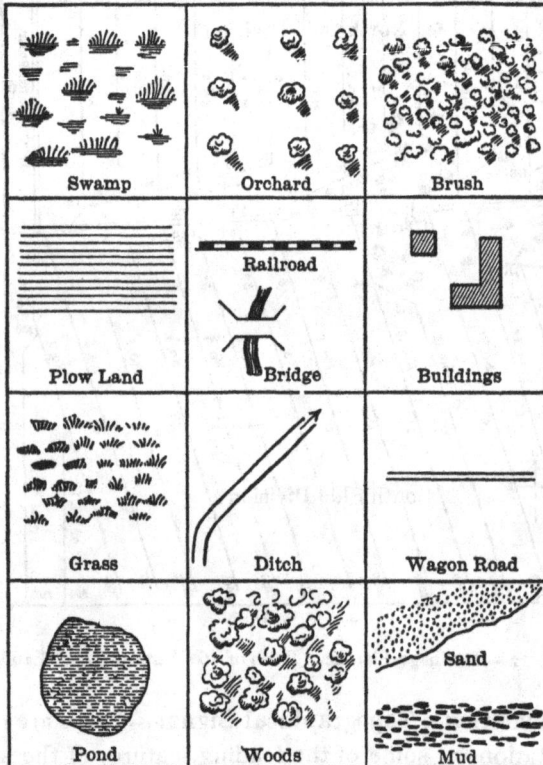

FIG. 23.—Conventional Signs used to Represent Topography.

Lettering.—The style in which the lettering of the map is done has a striking effect upon its general appearance.   When a working map is made, the lettering should be done in some free-hand letter which can be rapidly made with a writing-pen. The points that should be aimed at in this letter should be plainness, neatness, and such variety as will give a general air of finish to the work. Not very much time can profitably be given to this part of a working map, so that neatness and despatch in making it will be found of great advantage to the young engineer. The finished map should receive more careful attention in the execution of letters, particular attention being given to the adaption of the style of the letter to the importance of the thing for which it stands. Simple letters are best for drainage maps, as well as the most easily made. They should be lightly pencilled in until a design is found which is adapted to the map, after which they should be inked in. The letters should be the last part of the map to be finished and should be so placed as not to obscure figures which accompany lines.

Copying Maps.

It is sometimes desirable to make one or more copies of a map, either for use in the construction of the work or for preservation with the field notes. This is most conveniently done upon tracing-cloth or vellum. Stretch the cloth over the map and draw in the lines by letting the pen follow directly over the lines on the map. Instead of the tracing-cloth, a thin quality of bond paper may be used. This, however, is not so transparent as the cloth and hence requires a strong light in order that the lines may be readily traced.

Where several duplicates are required, what is known as blue prints can be made more cheaply than hand

tracings.   They are made as follows: Make a copy of
the map it is desired to print upon tracing-cloth or
upon a transparent paper, taking care to make all of the
lines definite and black.   Place this copy with its face
side against the glass of the print frame.   This frame
may be made of a strong picture-frame of sufficient
size to take in the drawing.   The back should be made
firm and so arranged that it can be clamped closely to
the glass by means of buttons attached to the back of
the frame.   Upon the drawing or map place a piece of
the prepared paper, the prepared face against the tracing, and place some smooth papers or layers of cloth
upon the back of it and clamp the backing board firmly
against it, so that every part of the tracing will be
firmly pressed against the glass.   The prepared paper
can be bought of dealers in engineers' supplies.   It
should be kept in a dry and perfectly dark place.
When both tracing and paper have been arranged in
the frame as above directed, expose the glass to the
direct rays of the sun for four or five minutes if the sun
is bright.   Take out the paper and place it in a bath of
clear water.   Move the water over it until the paper
turns a clear blue and the lines show a clear white.
Then hang up to dry.   If the blue is too light, the exposure was too short; if too dark, it was out too long.
If paper which is not fully transparent is used as a
copy it must be exposed longer than a perfectly transparent negative.   A little practice will soon enable the
novice to make good prints.

CHAPTER VIII.

GRADING THE DITCHES FOR TILE.

THE engineer should be thoroughly conversant with practical tile-laying and with the best plans for securing the most accurate work. Workmen can frequently dig the ditch, but do not understand the principles of grading it accurately according to survey. Many workmen think that it is more difficult to do than it is, hence the engineer should seek to make the whole matter plain. A survey should be carefully worked to or its full value cannot be realized. There are two good plans for grading a tile ditch, both of which depend upon the same principle and both simple and practical.

The Line Method.—This consists of setting a line or wire directly over the grade stakes at a given distance above and parallel to the bottom of the proposed ditch. As the bottom is finished for the tile it is tested by means of a gauge which carries a light crossbar set at a right angle to it. The line is stretched parallel to the grade line of the ditch and 5 feet above it, which is a convenient height, and tested by the gauge which is 5 feet long from the bottom to crossbar. The line should be supported at two or three points between stations to prevent sagging.

To set the line, subtract the depth of the ditch at a given station from the length of the gauge to be used and set the line above the grade stake the amount of this difference.   Then the distance below the hub plus the distance above it to the line equals the length of gauge.   This plan is illustrated in Fig. 24.

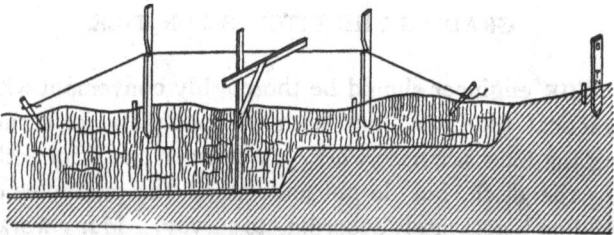

FIG. 24.—Grading by Gauge and Line.

Another method of finding the point at which to set the line does away with all mental subtraction of figures and the errors which may arise from it.   Take a stick the length of the proposed gauge,—in the above case 5 feet,—and graduate it to inches and quarter-inches, beginning at the top and numbering down. One-eighth inches can be obtained by estimation.   To use the measure at any grade peg, note the cut or depth for that stake, find the same mark on the measure, set the bottom end of the measure upon the grade peg and bring the line to this point.   When the measure is placed upon any grade stake, the position for the line is at the mark corresponding to the depth mark at that stake.

The Target Method.—Another method of grading called the target plan is better for large and deep ditches, and is in favor with many workmen for grad-

A FOUR-INCH LATERAL DRAIN.

(*To face page* 104.)

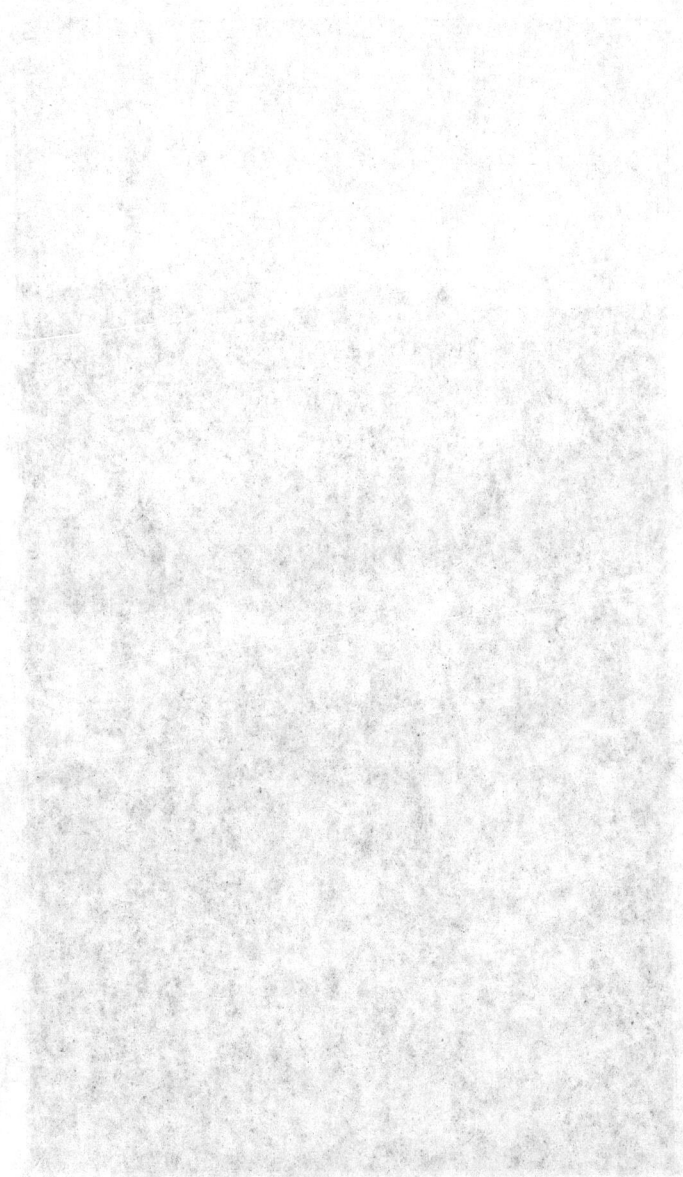

ing all kinds of drains.   It depends upon the same
principle as that described for setting the line, and consists of setting crossbars at the stakes instead of a line
and then testing the bottom by a line of sight over a
rod of the same length as that at which the bars are
set above the grade line.   Figure 25 will make the

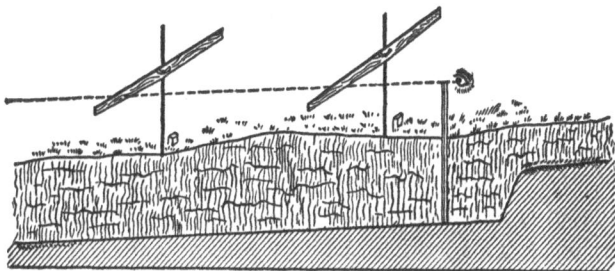

FIG. 25.—Grading by Sight Line and Target.

manner of using them plain.   The targets, as they are
called, are bars of wood about 3 feet long, and are
attached to an iron rod by means of adjustable clamps,
each of which is made tight by a thumb-nut.   The
target can be moved up or down upon the rod or set
at any angle desired.   The iron standard can be thrust
into the ground where it will remain firm, and the head
levelled and brought to the required height, at which
point it can be made fast by the thumb-nut.   When
two of these are set, the grade line can be worked to
in the manner shown in the cut.   One target should
be painted red, the other white, to aid in drawing a line
of sight over them.   Whenever there is a change of
grade the target should be reset, and at such times a
third target is necessary.

The only objection to this method is that some workmen do not have "a good eye" for this kind of work and cannot use it.

These methods of grading ditches are simple, practical, and accurate, and in the hands of competent workmen drains can be constructed with as great accuracy as our city sewers are laid when done under the most rigid supervision.

Digging the Ditches and Laying the Tile.

Not every laborer can dig a creditable ditch for tile, but perhaps any man can learn. An apprenticeship of greater or less duration is required to develop a skilful ditcher. This work is rapidly passing into the hands of those who by their skill merit a premium in wages readily accorded them by those who appreciate thorough and economical work.

The work of constructing a tile drain should begin at the outlet, be it a main or branch. The general method that should be followed may be described, but only practice will give that swing and ease of motion with which the trained workmen digs the ditch and lays the tile. The ditch should be started straight at the top and the curves should be smooth, not uneven and crooked. Let no one think that he can dig such a ditch without first drawing a line for cutting one side of it. This should be a $\frac{1}{4}$-inch rope which can be drawn tight or be laid to form a suitable curve. The top width of the ditch should be proportioned to the depth to which it must be made, 10 or 12 inches being a common width for a 3-foot lateral ditch. A ditching spade with blade 18 or 20 inches long, slightly curved forward and straight across the cutting-edge, is used for this work, the round-pointed finishing spade having been superseded by this.   The workman should thrust the blade into the earth a little quartering to the direction of the line, and take up nearly all of the earth cut loose on what is called the first spading. The round-pointed shovel with the long handle should now be used to throw out the loose earth.   The grading line having now been set up as previously described, the second or finishing spading is taken out.   The spade should be thrust down to within about 2 inches of the bottom of the ditch, and when about 4 feet in length has been excavated, the cleaning scoop should be used to remove the loose earth, and to cut a curved channel just large enough to receive the tile, the workmen all the time standing on the bench above the bottom.   The grade should be tested in the way heretofore described, and should not be passed until the few feet prepared is a perfect section of the continuous line as laid out.   If the ditch should be deeper than two spadings, as of course it frequently is, enough of the top should be taken off to permit the grade line to be reached by two full spadings.   In case of larger and deeper ditches the top width must be increased and also the number of spadings, but the process of taking the last one and finishing the bottom is the same.

The tile from 6 inches and smaller can be laid with a tile-hook if the bottom has been properly prepared (Fig. 26).   They may be turned until the ends fit together and lie firmly in the channel, which has been prepared for them by the scoop.   When finished, this line should have no crooks either in alignment or

grade.   The curves should be long, so that by turning
the tile in their beds close joints will be made.   If an
open joint is necessary on the outer side of the curve,

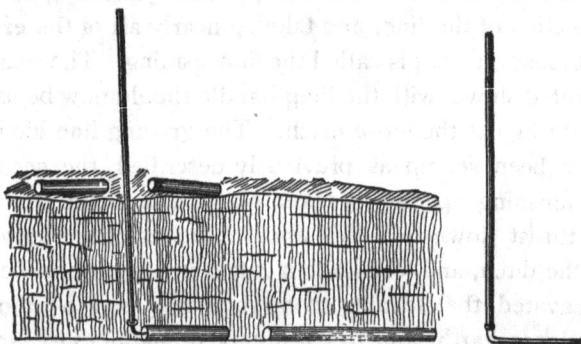

FIG. 26.—Method of Using Tile-hook.   Tile-hook.

a bat can be laid over it and clay tamped against it in
such a way that it will be secure.   All cracks that are
larger than $\frac{1}{4}$ inch should be covered with bats.   Following this method the drain is finished and secured as
the work proceeds from the outlet up grade.   When
the point for joining a branch is reached, the proper
tile with Y branch should be placed in position and
the opening covered securely to await the construction of the branch line.

When large tiles are used—9-inch and larger—it
will be necessary for the workman to walk in the bottom as he grades, which work must be done with the
shovel.   The same care should be used in securing the
proper grade as above noted.   Tile should be laid from
the outlet up, the workman standing in the bottom and
placing them with his hands.   They should be secured

in place by clay filling, which should be tamped firmly between the sides of the ditch and the tile. If the clay is so hard that it must be loosened by the pick, the tediousness and expense of the work will be greatly increased.

It will be observed that the tools necessary for this work are few: the line or target for obtaining the grade as given by the survey, a working line 100 feet long, ditching spade, round-pointed shovel, tile-hook, and cleaning scoop of the size required for the tile to be laid. The practice of many ditchers is to lay the tile by hand, walking backward in the ditch in front of the tile as they are laid. But it is wholly practicable, when the ditch is properly prepared at the bottom, to lay the tile with a hook from the surface in a perfectly, satisfactory way where the ditches are only 3 or 4 feet deep.

Difficulties in Constructing Tile Drains.

The engineer is often consulted regarding difficulties which are encountered in laying tile, and in his work as superintendent he is charged with the duty of helping out the contractor when he meets difficulties.

Of all difficulties which are encountered in constructing tile drains, quicksand or anything that resembles it in behavior is the most formidable to overcome. The ingenuity of the engineer as well as the skill of the vorkman is often taxed to the utmost in such cases. If great expense is to be avoided, probably the most sensible plan is first to select a dry season of the year in which to dig through soil known to contain quicksand.   Second, lay the drain as far into the treacherous soil as can be done safely, and then stop the work for a time until the water drains out to some extent and then proceed.   It may take a month or more to pass through a bad place, but it will be safer and cheaper than to attempt to force through by the use of sheeting or boxing.   As an aid to the solidifying of the mass so that it can be worked, temporary drains may be laid as far as possible and above grade in order to more rapidly draw off the surplus water.

To prevent the sand from entering the joints of the tile either tarred paper or coarse hay or grass placed closely about the joints has served the purpose, care being taken in all cases to lay the tile closely together. It is highly important in handling quicksand that the workmen should not disturb the material more than is absolutely necessary.   Each shovelful should be lifted carefully and without moving the adjoining sand. When once worked up into a thin mortar it cannot be handled except by baling.

Cleaning Tile Drains. — Notwithstanding that all possible care has been taken to prevent mud and sand from entering tiles during the construction of the drain, it frequently occurs that they will be found more or less obstructed from this cause.   If the tiles are in the required position, and are all right with the exception of the obstruction, do not disturb them, but clean out the mud or sand by the following plan: Remove the earth from 3 feet of the drain at intervals of about 20 feet.   Remove these 3 feet of tile and take out all of the material that can be reached.   Tie up a bundle of straw in a sack of such size that it will nearly fill

the bore of the drain.   Attach a rope securely to this
and pass the rope through the drain from one opening to the other.   This can be done by means of a
light flexible pole.   By means of this rope pull the
swab through the drain and as the material is forced to
the opposite end let it be dipped or shovelled out.   It
is well to have two ropes attached to the swab so that
having passed it through once it can be drawn back
and the operation reversed.   All this should be done
when there is but little water flowing through the drain.
After the stretch of drain which is obstructed is cleaned
out return the tiles which were removed to their original position.   A little mud or sand will always remain
in the drain after it has been scoured in this way, but
it will be readily washed out when the drain is flushed,
provided the drain is otherwise in perfect condition.

Submerged Outlet.—Where a submerged outlet is
necessary the drain must be laid when the ground is
dry or nearly so.   A submerged outlet in itself is not
objectionable, but it should be understood that the fall
or effective head of the drain is diminished by the depth
at which the water must rise above the outlet before
flowing away, and the line of soil saturation will extend back on a level until it intersects the line of the
drain.   With a proper head no injury will be done to
the drain at the outlet, and the rate of discharge will
be nearly as great, taking into account the diminished
head, as though the discharge were into the open air.
Submerged outlets are frequently a necessity in the
drainage of level tracts into artificial outlets, for the
reason that the drains must be placed so low with reference to the outlet channel that the tile outlets are

flooded at every considerable rise of water in the channel into which they discharge. This difficulty should
not deter one from laying drains under such conditions
provided the water in the open channel recedes quickly.
Where the drain for some distance back has a pressure of water or soil upon it, its action is similar to
that of an iron pipe, the head of drainage-water from
the upper levels being the force which causes the submerged outlet to discharge. The objection to submerged outlets is less for open soils than for retentive
ones, since in open soils additional head is given to the
drain when the soil in the various parts of the tract becomes fully saturated, or in other words the line of
saturation rises considerably above the floor of the
drains. Where the soil is retentive, the weight of the
soil water is but little more than will overcome the resistance which the soil particles offer.

Inspecting Tile Drains.

The most definite and satisfactory way of determining whether a drain has been laid as indicated by the
survey or not is to run over the work with the level.
Determine whether the tile is in the correct position
at the outlet point by taking a rod-reading on its grade
stake or some bench, by means of which the line may
be reproduced. Let the rodman rod up on top of the
tile at each station and also between each station and
at curves. The levelman records each reading, observing whether the differences correspond with the grade
as laid out, allowing in all cases margin enough to
cover the inequalities of the tile that have been used.

The rodman should observe the joints and whether improper tile have been used.   If the line " passes " the tile are ready to be blinded and the ditch filled.   If a line has been constructed skilfully $\frac{1}{4}$ inch per station should cover its variation from a true grade.   The judgment of the inspecting engineer should be somewhat carefully exercised in deciding whether certain faults that may be found will effect the efficiency of the work.   He must, however, have nerve enough to correct faulty construction and insist upon its being made right.   This is only justice to the employer and works no hardship upon the contractor, if what was expected of him was fairly set forth in the contract.

Heavy Rains on Unfinished Drains.—During the construction of a drainage system the work is often hindered in the spring of the year by heavy rains which fill to a greater or less extent the trenches which have been dug and submerge the lines of tile which have been laid.   In the case of mains with light fall, there is considerable risk from dirt and silt which may be washed into the drain and partially obstruct it.   The tile drain may be securely closed at the upper end, but if the water is permitted to flow over the top of the tile, the drain itself being nearly empty, the weight of the water passing down through the joints until the drain is full carries with it a large quantity of earth which by reason of the lack of current may not pass on through the drain.   The better way is to permit the water to enter direct through the end of the drain and fill the tile completely, letting the surplus pass on over the top of the drain.   The drain being full and the water flowing under a good head will prevent the top water from carrying silt into the drains.   Small drains, however, should be securely plugged up at the upper end.   Drains which in some way have become partially obstructed during construction may be cleaned in this way provided the required quantity of water is at hand at the right time.   It is desirable to fill the trenches as soon after the tile are laid as practicable in order to lessen the risk from injury by freshets.

Filling the Ditches.

Enough earth should be thrown upon the tile after they have been laid and inspected to secure them in their position.   This should be done by a careful workman, who should see that the earth is thrown around and over the tiles in such a way that they will not be moved by any subsequent filling.   This is practically the final inspection of the most permanent and lasting improvement which can be made in a soil.   Where the drains are in cultivated land, filling may be completed with a plough drawn by a team on each side of the ditch. The evener used on this plough for this work should be about 16 feet long.   It is assumed that the excavated earth has been thrown in about equal quantities on each side of the ditch.   If the land through which the drains pass is in grass, a V-shaped scraper made for the purpose and drawn with the point behind should be used, as this will move the earth without injuring the sod. Care should be taken to place all of the excavated earth in and directly over the ditch.   If this is not done, there will be a depression over every line when the loose earth settles firmly in place.   The fact that

good cement mortar.   A form is shown in Fig. 27.
Probably the most practicable outlet for farm drains,

FIG. 27.—Section of Stone Abutment for Drain Outlet.

especially where ditch banks are subject to shifting, is
the box with wire bars at the end, as shown in Fig. 28.

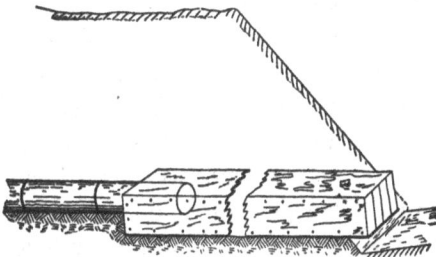

FIG. 28.—Box Outlets for Tile Drains.

The box should be not less than 12 feet long and set
on the grade line of the drain.   White-oak planks 2
inches thick should be used if they can be had, otherwise pine or hemlock.   Large galvanized iron wires
secured in a vertical position about 2 inches apart across
the end of the box will make a serviceable screen.

The connection of the box with the end of the line of tile should be thoroughly and carefully made, otherwise drainage-water will work its way around the box and underwash it. The earth should be most thoroughly tamped about the box as the filling is thrown in; in no case should it be loosely filled, as is done with ordinary ditches.

The Silt-basin.

Silt-basins are often injudiciously used in a drainage system. They serve two purposes: one to collect the water of several drains of a system so that it can be combined and discharged through a common main, and another to arrest the silt which a drain carries and cause it to deposit itself in the bottom of the basin or well from which it may be removed.

The effect of the silt-basin is to retard the flow of the water, and hence should never be used where this will prove injurious to the drains. The breaking of the current and the entrance friction of the outlet pipe, while it causes the silt to be deposited, also diminishes the velocity and discharge of the main drain, and is detrimental to the action of the drain on level grades. Where there is abundant fall and the entering drains can discharge into the basin not less than one foot above the top of the outlet drain it may be used.

The best plan for collecting the water of several drains into one is by means of the Y junction and a "drop" from branch into main as noted heretofore. The current is preserved, and silt is carried to the outlet before it is deposited. Where the drains are well laid

and the subsoil is clay, there is no occasion for the use
of the silt-basin in land drainage systems. Its use for
a catch-basin for conveying surface water into tile
drains under certain conditions will be described hereafter.

Digging Tile Ditches with Machines.

Many machines of different patterns have been pat-
ented, tried, and have failed to displace hand labor in
digging tile ditches. The difficulties to overcome do
not seem insurmountable, nevertheless the history of
such machines discloses partial successes which for a
time promised well, but in the end did not meet the requirements of the work in all kinds and conditions of
soils. Ploughs of different patterns have been used to
aid the spade by loosening the earth in the bottom of
the ditch, thereby diminishing the labor of excavating
somewhat. The machines which have dug the ditch
to its full depth at one passage have given the most
satisfactory results. The ditch is left completed to
grade and ready for the laying of the tile, the excavated earth is left loose and ready to be easily back-
filled and the grade is more easily and accurately made
than by a machine which excavates by many passages
over the ground.

One of these machines which gives excellent results
at the present writing and has been more successfully
introduced than its predecessors is distinctively a
traction ditcher. It is propelled by steam power
which operates a cutting wheel and at the same time
moves the machine forward by traction, excavating a

completely graded ditch at the rate of from 10 to 20
rods per hour. It is governed and operated in the
same manner as a traction engine, is compactly built
and easily managed by two men. It cannot be expected that a machine of this character will work successfully in ground so soft that it will not bear the
weight of the machine, nor in land full of stumps,
stones, and large roots.

With these exceptions, there seem to be no difficulties which are not successfully met by this machine.

Contracts for Construction of Tile Drains.

The construction of drains for improving entire farms
and large areas of level land is frequently done by contract subject to competent superintendence. The system is laid out by an engineer and complete plans made
for the work, after which bids for the construction of
the drains are solicited. The furnishing of the tile and
distribution as needed upon the ground is usually not
included in the ditching contract, neither is the filling
of the trenches.

Tile are purchased by the thousand feet; hauling
from the railroad station or factory is done by the ton,
the weight of the individual pieces of different sizes
being used as a basis for determining the weight of the
loads; the digging of ditches and laying of tile is commonly done by the rod, but by the 100 feet would be
better, and filling of the ditches by the 100 feet.

The following general specifications and contract
have been found useful in practice, and will serve as a
guide in preparing specifications for other work. It

will be observed that the only bond or security required of the contractor is the 25 per cent retained until the completion and acceptance of the work.   The only method of securing good work of this kind is to give it thorough supervision and know that it is right before the drain is covered.   When a drain has been accepted the liability of the contractor should cease provided the individual line has been finished.   The retaining of 25 per cent is made for the purpose of securing the correction of any faults in the work which may be discovered before the drain is covered.

GENERAL SPECIFICATIONS FOR THE CONSTRUCTION OF TILE DRAINS.

THE lines for the ditches are indicated on the field by stakes which have been set by the engineer, and the depths and grades given by him shall constitute a part of these specifications.

Digging the Ditches.—The digging of each ditch must begin at its outlet, or at its junction with another tile drain, and proceed toward its upper end.   The ditch must be dug along one side of the line of survey stakes, and about ten inches distant from it, in a straight and neat manner, and the top soil thrown on one side of the ditch and the clay on the other.   When a change in the direction of ditch is made, it must be done by means of a neat curve, but in all cases the ditch must be kept near enough to the stakes so that they can be used in grading the bottom.   In taking out the last draft, the blade of the spade must not go deeper than the proposed grade line or bed upon which the tiles are to rest.

Grading the Bottom.—The ditch must be dug to the depth indicated by the figures given with the survey, which depth is to be measured from the grade stakes which are set for that purpose, and graded evenly on the bottom by means of the " line and gauge " method, or " target," or any other equally accurate device for obtaining an even and true bottom upon which to lay the tile. The bottom must be dressed with the tile hoe, or in case of large tiles, with the shovel, so that a groove will be made to receive the tile, and when laid in it they will remain securely in place.

Laying the Tile.—The laying of the tile must begin at the lower end and proceed up-stream. The tile must be laid as closely as practicable, and in lines free from irregular crooks, the pieces being turned about until the upper edge closes, unless there is sand or fine silt which is likely to run into the tile, in which case the lower edge must be laid close, and the upper side covered with clay or other suitable material. When, in making turns, or by reason of irregular-shaped tile, a crack of one fourth inch or more is necessarily left, it must be securely covered with broken pieces of tile. Junctions with branch lines must be carefully and securely made.

Blinding the Tile.—After the tile have been laid and inspected by the person in charge of the work, they must be covered with clay to a depth of six inches, unless, in the judgment of the superintendent, the tile are sufficiently firm, so that complete filling of the ditch may be made directly upon the tile. In no case must

the tile be covered with sand without other material
being first used.

Risk During Construction.—The ditch contractor
must assume all risks from storms and caving in of
ditches, and when each drain is completed it must be
free from sand and mud before it will be received and
paid for in full.   In case it is found impracticable, by
reason of bad weather or unlooked-for trouble in digging the ditch, or properly laying the tile, to complete
the work at the time specified in the contract, the time
may be extended as may be mutually agreed upon by
employer and contractor.   The contractor shall use all
necessary precaution to secure his work from injury
while he is constructing the drain.

Tile to be Used.—Tile will be delivered on the
ground convenient for the use of the contractor.   No
tile must be laid which are broken, or soft, or so badly
out of shape that they cannot be well laid and make a
good and satisfactory drain.

Payments for Work.—Unless otherwise hereafter
agreed upon, the contractor may at any time claim and
receive from the employer seventy-five per cent of the
value of completed and accepted work at the price
agreed upon in the contract.   Twenty-five per cent
will be retained until the entire work contracted for
is completed and accepted, at which time the whole
amount due will be paid.

Prosecution of the Work.—The work must be pushed
as fast as will be consistent with economy and good
workmanship, and must not be left by the contractor
for the purpose of working upon other contracts, except by permission and consent of the employer.   All

survey stakes shall be preserved and every means taken to do the work in a first-class manner.

Failure to Comply With Specifications.—In case the contractor shall fail to comply with the specifications, or refuse to correct faults in the work as soon as they are pointed out by the person in charge, the employer may declare the contract void, and the contractor, upon receiving seventy-five per cent of the value of completed drains at the price agreed upon, shall release the work and the employer may let it to other parties.

Sub-letting Work.—The contractor shall not sublet any part of the work in such a way that he does not remain personally responsible, nor will any other party be recognized in the payment for work.

Plans and Tools.—The contractor shall furnish all tools which are necessary to be used in digging the ditches, grading the bottom, and laying the tile. In case it is necessary to use curbing for the ditches, or outside material for covering the tile where sand or slush is encountered, the employer shall furnish the same upon the ground convenient for use.

All plans and figures furnished by the engineer, together with the drawings and explanations, shall be considered a part of these specifications.

CONTRACT.

It is hereby agreed between........................ ....
employer, and ... ., contractor,
that the said .... will construct the following named or described tile drains in accordance with the foregoing
specifications, at the prices herein named, and that he will begin the
work on or before............and complete the same by.............

....
....
....
....
....
....
....
....
....
....
....
....

Witness the hands of the respective parties, this..............day of
....

........................., Employer.
...., Contractor.

CHAPTER IX.

FLOW OF WATER THROUGH PIPES.

BEFORE taking up the subject of the size of drains it is proposed to outline briefly that branch of hydraulics which relates to the flow of water through pipes, in order that the beginning engineer may have the basis upon which to study the flow of water through tile drains. While there have been very elaborate investigations made upon this subject, it is only necessary in this connection to outline the facts without entering into a full analytical discussion of the matter.

It should be borne in mind by those who are planning drainage works that gravity is the sole cause of flow in water. The flow of water in a drain of whatever kind is owing to the inclination of the surface of the fluid. The same universal force causes unsupported bodies to fall vertically, a ball to roll down an incline, and water to flow along a channel or through a pipe. The formula used to express the theoretical velocity due to gravity is

$$v = \sqrt{2gh},$$

where v = velocity in feet per second;

g = accelerating force of gravity;

h = space through which the body falls.

It has been found by experiment that a body in vacuum at the level of the sea passes through a space of

16.1 feet during the first second, and at the end of that time has acquired a velocity of 32.2 feet. The velocity at the end of each succeeding second of time is 32.2 feet greater than it was at the end of the preceding second. This is called the accelerating force of gravity and is usually designated by g. Knowing the height that the body has fallen at any time, the velocity may be determined. The following table shows at once the relation of time, space, velocity, and accelerating force of the action of gravity upon falling bodies during the first five seconds:

FALLING BODIES DURING FIRST FIVE SECONDS.

Time...	1 sec.	2 sec.	3 sec.	4 sec.	5 sec.
Space = h............	16.1	64.4	144.9	257.6	402.5
Velocity = v..........	32.2	64.4	96.6	128.8	161.0
Accel. force = g......	32.2	32.2	32.2	32.2	32.2

Water flowing down an inclined surface would follow the same law were it not for resistances of various kinds which constantly act upon the particles of water as they descend and produce a uniform flow where the resistance is constant. Were this not the case, our ponds and lakes would soon become dry, and our brooks and rivers would for a time become dangerous torrents. The equilibrium of natural forces would at once become unbalanced.

It has been the life-long work of many eminent hydraulicians to determine by practical experiments the value of these retarding forces and by introducing them into the gravity formula to so modify it that it shall be a correct expression for the flow of water under given conditions and thus become of use in practical affairs.

Simple as this task may at first seem, it has occupied
the time and attention of many experimenters who are
justly noted by reason of their researches in this department of practical science.

When the flow of water through pipes is considered,
the resistances to gravity are, first, resistance to entrance of water into the pipe; second, the resistance
offered by the walls of the pipe with which the water
comes in contact. The first will vary with the kind of
opening through which the water enters the pipe, the
second with the kind of pipe and its length and the
head which produces flow.

Many of the formulas deduced agree with each other
in the results obtained from them sufficiently near for
practical purposes, yet there is a wide difference in their
simplicity and availability for use. There is one which
represents at a glance the corrections that must be applied to the gravity formula, so that it will express the
velocity of water in pipes. It is known as Weisbach's
formula and is expressed as follows:

$$v = \frac{\sqrt{2gh}}{\sqrt{e + c \times \dfrac{l}{d}}}, \quad \cdots \quad (2)$$

where e = coefficient of resistance to entrance of water
into pipe;

c = coefficient of friction of pipe;
l = length of pipe in feet;
d = diameter of pipe in feet.

The numerator of the second number of this equation
is the theoretical velocity of falling bodies. The de-

nominator represents the resistance offered by the pipe to the flow of water through it as determined by the author of this formula. When the numerical values found for e and c are substituted, the result is the velocity formula for falling bodies modified so that it will apply to the velocity of water in pipes.

There are formulas which are just as accurate and possess the very desirable excellence of greater simplicity in which g and h are combined with quantities whose value has been determined by experiment. Beardmore's formula is one of these simple expressions:

$$v = 100 \sqrt{RS}, \quad . \quad . \quad . \quad . \quad (3)$$

where $R =$ hydraulic mean depth

$$= \frac{\text{area of waterway}}{\text{wet perimeter}} = \frac{a}{p};$$

$$S = \text{sine of slope} = \frac{\text{head in feet}}{\text{length of pipe}} = \frac{h}{l}.$$

This formula written out in full is as follows:

$$v = 100 \sqrt{\frac{\text{area}}{\text{perimeter}} \times \frac{\text{head}}{\text{length of pipe}}}. \quad . \quad (4)$$

In this expression the constants g and coefficients of friction are merged into one common constant, 100, and the variable quantities are expressed in terms which may always be determined for each particular pipe.

The above are examples of reliable velocity formulas made use of by engineers in computing the flow of water through pipes of various kinds and sewers of various descriptions when the head of water is known and the pipes come within reasonable limits of perfection in workmanship. The American unit is feet per second and the quantity discharged is cubic feet per second. When the velocity is found the discharge is obtained by multiplying the area of the column or jet of water expressed in square feet by the velocity in feet per second. The result will be cubic feet per second. By formula the expression would be

$$Q = av, \quad \cdots \quad \cdots \quad (5)$$

where Q = quantity in cubic feet;

a = area of column of flowing water;

v = velocity as determined by formula.

Substituting the value of v in equation 5 we have

$$Q = 100a \sqrt{RS} \quad \cdots \quad (6)$$

Flow of Water through Tile Drains.

As previously stated, the object of draining land is the removal of such soil water as is not needed for the profitable growth of the plants we desire to produce. The source of all water is the rainfall as it is distributed over the surface of the soil at irregular times and in varying quantities. The removal of the part not wanted is accomplished by underground tile drains, the water reaching them by percolation through the soil. The problem in drainage hydraulics is not only to determine the quantity of the water which a certain drain will carry, but also to ascertain how much water should be removed from the soil at certain times in order to place the land in the desired condition.

The very quick and rapid removal of soil water is not desirable in the drainage of farm land. The object should be to remove the surface water quite quickly and secure a gradual movement of sub-surface water through the soil into the drains.   The action of the water in passing slowly through the soil is beneficial in imparting to it such fertilizing gases as it may have absorbed from the atmosphere, and in disintegrating soil particles and helping to prepare them for plant food.   With this object in view, sufficient drainage is better than too much.

The rainfall is exceedingly variable both with respect to the season of precipitation and the quantity which falls, so that a system of drains which would prove ample during one year, or even for a series of years, would at other times become overcharged and prove insufficient for the work desired.

The land to be drained is in some localities a level tract, in others it is broken up by ridges, slopes, ponds, and swamps alternating in irregular variety, thereby greatly complicating the plans which must be used for draining successfully as well as the determination of the size of the drains for the same.

The drains ordinarily used for the work are not uniform either in the quality of the material or excellence of the workmanship when constructed.   The drain receives water at all of the joints, and in the case of mains at various points where branches join.   These discharge into the main under varying heads and grades.

It follows that it is no easy matter to formulate the elements which enter into economical and efficient land drainage so as to put the subject of the flow of water in drains on a scientific and at the same time a practical basis.

It has been found, however, by observations and

experimental tests that the existing formulas for the flow of water through pipes are with some allowances more applicable to tile drains than at first would be thought possible. The art of using a formula successfully is in bringing to it in proper form the data which should be used and in making the proper assumptions which should precede the application of any formula.

The following formula has been found to be simple and of easy application and has been verified by use of a current meter upon the discharge from lines of tile in which the grade of the drain and the quality of workmanship in its construction were known:

$$\text{For velocity} \quad v = 48\sqrt{\frac{df}{l + 54d}}, \qquad (7)^*$$

$$\text{For discharge} \quad Q = 48a\sqrt{\frac{df}{l + 54d}}, \qquad (8)$$

where v = velocity in feet per second;
$\quad d$ = diameter of tile in feet;
$\quad f$ = total fall in length of drain in feet;
$\quad l$ = length of drain in feet;
$\quad a$ = area of tile in feet;
$\quad Q$ = discharge in cubic feet per second.

All measurements to be taken in feet.

Application of the Formula.

This formula will give quite correctly the velocity and discharge from lines of tile drains which are laid in a first-class manner, within certain limitations as to size and length, which will hereafter be considered. In ap-

---

* Known as Poncelet's formula.

plying it for determining the size of main drains the question arises how much water per acre should be taken off the land in a given time Many engineers who discuss this branch of the drainage subject propose some formula which is used for water supply or for city sewerage and by it compute the size a drain should be to carry off 1 inch or $\frac{1}{2}$ inch of water in twenty-four hours These results differ so widely from the most approved practice in drainage that they are in disrepute among practical men and are not sustained by the test of experience. The discrepancy is explained by saying that drainage is not thorough as practised. Be that as it may, observation and experiment demonstrate quite conclusively that where thorough work is practised, tile drains do not remove $\frac{1}{2}$ inch of water from the soil in twenty-four hours. While this standard is about correct for open ditches, it does not apply to ordinary tile drainage, except under special conditions.

The results of experiments and observations in recent years seem to sustain the practice of many careful engineers who state that the removal of $\frac{1}{4}$ inch depth of water in twenty-fours hours by tile drains meets the requirements of the average soil, and may be regarded as a basis upon which to make computations for the size of mains, subject to such changes as the special tract to be drained may demand. The following are some of the considerations which have a bearing upon this question.

A drained soil is a reservoir. The ideal drained field is one in which water sinks downward through the soil from the point where it falls, the surplus passing into some drain near by and thence into the main. If

the drains are laid 3 feet deep we have a soil reservoir of that depth which, if it is very dry, will hold
one third of its depth in water.   But this condition is
rarely if ever met with in practice, but if the soil is
well drained it will take 2 inches of rain to saturate it,
the amount, however, depending upon the soil.   The
evaporation, which in the summer begins at once, and
also the absorption and evaporation of moisture by
plants, make a large draft upon water of the soil.

Most of the data on evaporation and drainage are
taken from observations made in England.   This does
not apply to our soil and climate for the reason that
the atmosphere of England is much more humid and
the evaporation from plants and soil much less than it
is in this country.   It has been shown by good authority
that evaporation from soil and water in this country
is fully twice as great as that in England, and it follows that a like difference should exist in evaporation
from plants.

An experiment made in 1889 at Uniontown, Ala.,
for the purpose of finding the percentage of the rainfall
which passes off through tile drains is stated briefly as
follows:

The tract of land consisted of three acres drained by
lines of tile 3 feet deep and 30 feet apart.   The measurements recorded were the outflow of excessive rainfall
April 13th and 14th, when 3.39 inches fell in twenty-four
hours upon land denuded of vegetation by the preparation for spring crops.   From the measurements made it
appears that the greatest discharge from the drains was
about eighteen hours after the first heavy rainfall, at
which time the drains were discharging $\frac{1}{8}$ of an inch

in twenty-four hours.   At the end of nine days only 23 per cent of the rainfall had passed off in drainage.

It must be admitted that there is a wide range of differences in the recorded results of experiments made for the purpose of determining the percentage of drainage to rainfall.   One series of carefully constructed observations gives the average filtration for a series of years at 5 per cent of the rainfall.   Another series apparently as reliable gives 42 per cent.   All of them, however, give a small percentage for the months of May, June, July, and August.   The above observations were not made in this country, and furthermore the means which were used in arriving at these conclusions were crude.

Soil water should occupy from 10 to 30 per cent of the empty space in productive soils.   Not until the quantity exceeds this will the drains be called into action. Again the subsoil for a foot or more above the plane of the lateral drains may be saturated for a time and no injury result to the upper soil or the growth upon it. One benefit resulting from a system of underdrains which is often overlooked is that soil water even when excessive does not stagnate.   It moves constantly in the direction of the drains, and air fills the space formerly occupied by surplus water.

There is another fact bearing on this question which should be considered in this connection.   Owing to the rapidity of rainfall on some occasions, and the inability of some soils to absorb the water quickly enough, a certain portion of it flows over the surface and collects in depressions near by and must be removed from those points either through the underdrains or by some surface relief.   Such conditions give rise to the complaint that the drains are not sufficiently large. Additional drainage should be provided at such depressions either by the use of more than the usual number of underdrains or by some surface overflow. It is always wise to place the drains closer together in the depressions than on the surrounding land, not that more water is precipitated there, but they are the natural receptacles for surface overflow. Every means possible should be taken to intercept overflow water from the higher lands.   The actual head under which a drain works when discharging its maximum quantity is the difference in elevation of its outlet and head end, to which should be added at least one half of the depth of the drain at the upper end.   A porous soil filled with water nearly to the surface easily gives this additional head and should be used in the formula when computing the full capacity of the drain.   It is the impression among many engineers that a city pipe sewer forms a more perfect channel for the flow of water than a tile drain.   Observations upon this point indicate that this opinion is unfounded.   A well-laid tile drain is graded as accurately as the best of sewers; there is less resistance at the joints, for the cemented joints in a pipe sewer are rarely smooth and perfect; the flow of a sewer must carry more or less solid matter, and the surface of the pipe is often coated with a slime which retards flow, while on the other hand the water in a tile drain is clear and the walls of the tile are clean.

Keeping in mind the cautions and conditions heretofore enjoined, formula 7 may be applied for determining the size of mains and submains.   Assume a size

TABLE 2.

AREAS OF TILE IN SQUARE FEET

Dia. n Inches.	Dia. in Feet.	Area in Sq. Ft.	Dia in Inches	Dia. in Feet	Area in Sq Ft
2	.1667	.0218	11	.9167	.6600
3	.2500	.0491	12	1.000	.7854
3½	.2917	.0668	13	1.083	.9218
4	.3333	.0873	14	1.167	1.069
5	.4167	.1363	15	1.250	1.227
6	.5000	.1964	16	1.333	1.396
7	.5833	.2673	17	1.417	1.576
8	.6667	.3491	18	1.500	1.767
9	.7500	.4418	19	1.583	1.969
10	.8333	.5454	20	1.667	2.182

TABLE 3.

HEAD IN INCHES PER 100 FT. REDUCED TO FEET PER 100 FT AND FEET PER MILE.

Head in Ins per 100 Ft.	Head in Ft. per 100 Ft.	Head in Ft. per Mile
	.0052	.274
	.0104	.549
	.0208	1.098
	.0313	1.652
	.0417	2.201
	.0521	2.750
	.0625	3.300
	.0729	3.849
1	.0833	4.398
	.0938	4.952
	.1042	5.501
	.1146	6.050
	.1250	6.600
	.1354	7.149
	.1458	7.698
	.1563	8.252
2	.1667	8.801
	.1771	9.350
	.1875	9.900
	.1979	10.449
	.2083	10.998
	.2188	11.552
	.2292	12.101
	.2396	12.650
3	.2500	13.200
	.2604	13.749
	.2708	14.298
	.2813	14.852
	.2917	15.401
	.3020	15.950
	.3125	16.500
	.3229	17.049
4	.3333	17.598

TABLE 4.

CONTENTS OF TILE IN ONE FOOT OF LENGTH

Dia in Inches.	Cu Ft in 1 Ft Length.	Gallons of 231 Cu Ins in 1 Ft Length
2	.0218	.1632
2½	.0341	.2550
3	.0491	.3672
3½	.0668	.4908
4	.0873	.6528
4½	.1104	.8263
5	.1364	1.020
5½	.1650	1.234
6	.1964	1.469
7	.2673	1.999
8	.3491	2.611
9	.4418	3.305
10	.5454	4.080
11	.6600	4.937
12	.7854	5.875
13	.9218	6.895
14	1.069	7.997
15	1.227	9.180
16	1.396	10 44
17	1.576	11 79
18	1.767	13.22
19	1.969	14 73
20	2.182	16 32

and apply the formula, being careful to consider the physical conditions of the area for which the computation is made. Substitute the proper numbers in formula 7 and find the value of v; multiply the value of v by the area of the tile as found in Table 2 to find value of Q. The result will be the quantity of water in cubic feet per second which the tile will discharge. Divide this number by the number in Table 5, which expresses the quantity of rainfall per acre which it is desired to remove per second in twenty-four hours. The result will be the number of acres which the given drain will afford an outlet for.

TABLE 5.

TABLE OF CUBIC FEET PER SECOND WHICH MUST BE DIS-CHARGED FROM A DRAIN TO RELIEVE ONE ACRE OF LAND OF VARIOUS DEPTHS OF WATER IN 24 HOURS.

.0420 cu. ft..	1 inch per acre
.0315 " "	$\frac{3}{4}$ " " "
.0210 " "	$\frac{1}{2}$ " " "
.0140 " "	$\frac{1}{3}$ " " "
.0105 " "	$\frac{1}{4}$ " " "
.0052 " "	$\frac{1}{8}$ " " "

EXAMPLE.—How many acres will a 6-inch tile drain, 1000 feet long, laid on a grade of 3 inches per 100 feet and 3 feet deep at the upper end, computing on the $\frac{1}{4}$-inch standard?

$$\text{Formula (7):} \quad v = 48 \sqrt{\frac{df}{l + 54d}} \qquad \begin{array}{l} d = .5; \\ f = 2.5 + 1.5 = 4; \\ l + 54d = 1027. \end{array}$$

$$\frac{2}{1027} = .00194$$

$$48 \sqrt{.00194} = 2.112 = v.$$

$$Q = av = 2.112 \times .1964 = .41479.$$

$$\text{Acres} = \frac{.41479}{.0105} = 39.5.$$

NOTE.—Add one half the depth of drain at upper end = 1.5 feet to the fall for value of f.

How many acres when the tile is laid on a grade of 10 foot per 100 feet ?

$.5 \times 2.5 = 1.25 = df.$

$d = .5;$
$f = 1 + 1.5 = 2.5;$
$l + 54d = 1027.$

$$\frac{1.25}{1027} = .00121; \quad \sqrt{.00121} = .0348; \quad 48\sqrt{.00121} = 1.67 = v.$$

$$Q = 1.67 \times .1964 = .32798.$$

$$\text{Acres} = \frac{.327988}{.0105} = 31.23.$$

NOTE.—Making f = grade only, which should be used where the soil is close.   Acres = 20.

To find the volume of water in cubic feet per second which a tile drain will discharge, multiply the computed velocity in feet per second for the diameter of tile required by the number given in column 3 of Table 2.

NOTE.—To reduce cubic feet to gallons multiply by 7.48.

Weight of 1 cubic foot of water........................... 62¼ lbs.
Weight of 1 gallon of water...............................8.35 lbs.
Seconds in 1 hour... 3600
Seconds in 24 hours.......................................86,400
Number of cubic feet of water on 1 acre of land when covered 1
  inch deep.. 3630

The Total Head or Fall that should be Used.—In considering the total fall of a main drain it is quite evident that the head which generates the flow is not in all cases the difference in elevation of the two ends of the line, but when supplied by a system of laterals which have a greater rate of incline the velocity head will be increased by the laterals.   Also, if the soil is saturated above the tile to nearly the surface, and the soil is free and open, the main drain as well as the laterals will receive added head from this source. While these forces are variable and difficult to formulate, it is safe to assume a small additional head made up from these sources.

To find the total flood head, find how much higher the upper terminal of each lateral or sub-main is than the upper terminal of the main drain, take the mean of the differences and add it to the head of the main. To this result add one half of the depth of the upper end of the main, for open soils only, for the value of f in the formula.

By Formula.

$$f = h + \frac{b}{n} + \tfrac{1}{2}C. \quad . \quad . \quad . \quad (9)$$

h = head of main.

b = sum of excess of elevation of upper terminals above upper terminal of main.

n = number of laterals having additional head.

C = depth of main at upper end.

Areas Drained by a Main.—The number of acres that a main is to actually drain is not the only element to be considered in this connection, but the manner in which the water is to be brought to the main, together with the general outline and contour of the surface, should also be carefully investigated. The formula is based upon the supposition that the surface of the tract is reasonably even, and that the drainage is brought to the main by lateral tile drains. Should a main be laid through a watercourse like a slough, into which there is considerable lateral surface drainage with no lines of tile, the main is at the disadvantage of receiving all of its water along one narrow line, instead of taking it as collected by a series of laterals and brought by them to various points along its course. The same is true of many drains which are laid into ponds. The pond collects surface water from a considerable area, and none of it can enter the drain until it reaches the center of the pond.

In such cases and many similar ones which occur in practice, the acreage should be converted to its equivalent of tile-drained land before the formula is applied for the purpose of determining the size of drain which should be used. Find the area of the land which has a surface slope in the direction of the drain, and which is not tiled, add one half to the area if in general it has a slope of 2 or 3 to 100, and apply the formula to the corrected acreage. If the slope is steeper, the acreage should be increased proportionally. For example, a pond through which a drain passes may have a water-shed of six acres, but may slope toward the center at the rate of 2 or 3 per 100. The drain should not be proportioned for six acres, but for one half more, or nine acres. Mains are not proportioned according to the number of lines of tile which are to discharge into them, but according to the area of land from which they must take the drainage.

Limitations of Size, Grade, and Length of Drain.

From what has preceded, it will have doubtless occurred to the reader that there are certain limitations to the size of the drain used, its grade, and the total length of the line, beyond which it is not wise to go. In formula No. 2 the denominator of the second number represents the resistances which oppose gravity.

If the sum of these equal gravity, no flow takes place. It may be observed in this connection, by way of explanation, that head is considered as divided into two parts. One part overcomes friction, and is called friction head, the other generates velocity and is called velocity head. The v of the formula is the mean velocity of actual flow from the tile which is under consideration, and is the difference between the friction head and the velocity head.

Considering the fact that friction is greater for small tiles than for large, and greater for long drains than for short ones, and that the velocity decreases as the square root of the head, it is easy to see that a drain of certain size may be laid with so little grade and have such a length that the discharge will be little or nothing. These facts will appear to some extent when the formula is used, yet it will be best to observe in a general way the following limits:

TABLE 6.

LIMIT OF SIZE OF TILE TO GRADE AND LENGTH.

Size of Tile in Inches.	Minimum Grade per 100 Feet.	Limit of Length in Feet.
2	.10	600
3	.09	800
4	.05	1600
5	.05	2000
6	.05	2500
7	.05	2800
8	.05	3000
9	.05	3500
10	.04	4000
11	.04	4500
12	.04	5300

This table means that the size of tile given should not be laid on a less grade nor in lines of greater length, when laid upon that grade, than is given opposite the size. If the formula should call for a certain size when the length is greater than this, then use a larger size than that denoted by the formula.

CHAPTER X.

SIZE OF LATERAL DRAINS.

THE size of tile that are to be used on a drainage system should be determined only after all of the data and knowledge to be obtained regarding the land is at hand and all of the grades have been figured out. The following hints as to the manner of taking up this division of the work may be observed with profit.

If there are both mains and sub-mains, find how much land is to drain through the outlet of the main, and also how great an area will be drained by each one of the sub-mains. Determine the size of each outlet by the formula, observing carefully the manner of using it. Diminish the size of these drains up-stream according to the area to be drained to that point, taking into account also the grades of the several parts. If the grade is light and uniform the large sizes should be continued well up towards the source of the drain. Each water-shed should be considered carefully by itself and the several basins be connected with each other by the main drain so that the whole will balance and work as one drain.

Size of Tile for Laterals.—No attempt should be made to make the capacity of intercepting drains equal the combined capacity of the laterals where a system of thorough drainage is employed The size of laterals

DIGGING A DITCH FOR TWELVE-INCH DRAIN-TILE.

(To face page 144.)

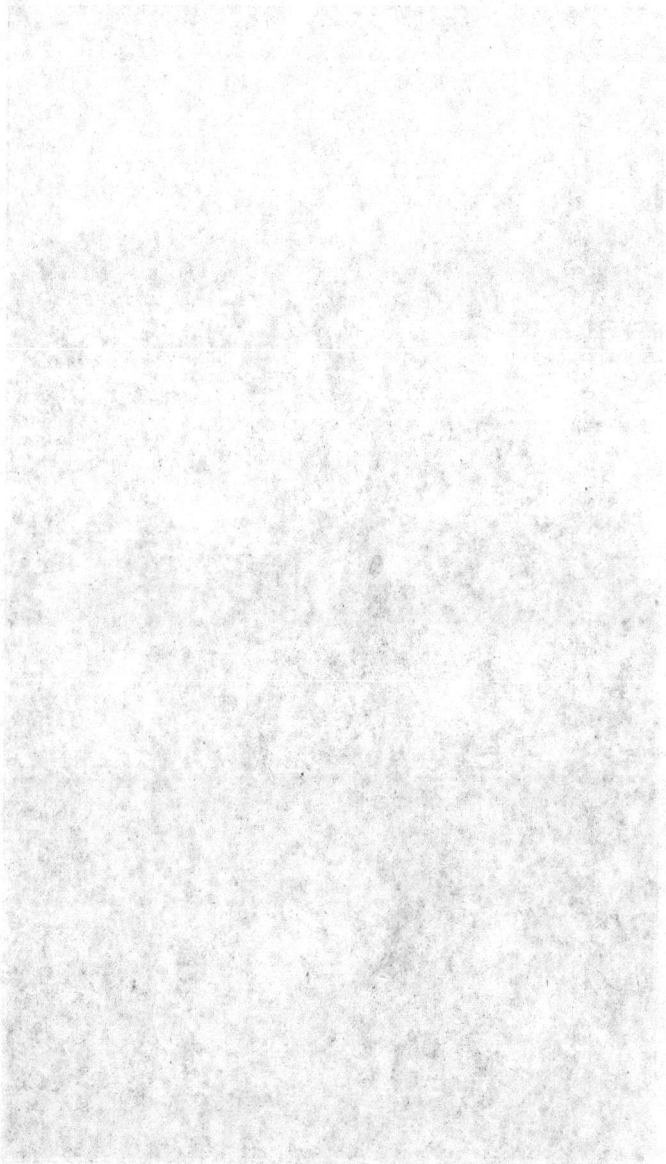

where the soils are open and permit the use of drains 100 or more feet apart should be 4- and 5-inch tile, which in some cases may be diminished to 3 inches at the upper end. For drains at a less distance apart 3-inch tile may be used for laterals. They are usually required to carry only a small part of their full capacity in order to relieve the soil of its surplus water. To do this well, however, they should not be quite full at any time unless it be when there is more than ordinary rainfall. They do not work under a pressure head, hence their velocity of flow is as great when running half full as when running full. Water in a drain attains greatest velocity when three fourths full and its greatest rate of discharge when nine tenths full.

Should the system of laterals have a heavy grade compared with the mains into which they discharge, the mains will often work under pressure and run full when the rainfall is not very large. They will of course run no more than full under a greater head, but the velocity and discharge will be greater. It is often thought that since a main drain is running full, it is doing all that it is capable of doing. From what has been said above, it will be seen that this is not the case.

Provision for Unusual Rainfall.

The rules thus far given for the size of tile do not provide for the carrying off, through the tile system alone, unusually heavy rainfalls of 2 to 3 inches in twenty-four hours. These occur sometimes two or three times a year. Again, two or three years may pass without witnessing such a downpour. It is still more rare

that these excesses come at a season of the year when a
short flooding will injure growing crops, yet these facts
should all be taken into account when we provide a
drainage system.   There are two difficulties which are
encountered in providing for floods with tile drains.
The first is that there are many soils that will not permit the water to pass through them rapidly enough
so that the water will not accumulate upon the surface
when the rainfall is sudden and large in quantity.   The
second is that the tile which should be used to carry it
all off must be of double the ordinary capacity, which
involves a great expense.

In general land drainage it will be found wise to provide for the excessive rainfall by keeping well-con-
nected depressions or broad shallow open ditches along
the course of natural drainage, so that the flood water
will pass off over the surface and thus relieve the under drains of the excess.   In level lands, where there
is the most necessity for this provision, no harm will
come from surface washing, for as soon as the excess
passes off over the surface the remainder of the drainage passes off through the soil.   There is no loss of
land, since in most cases these broad depressions or
ditches can be cultivated.

Where the fall is very considerable, the capacity of
the underdrains will be greatly increased by reason of
the grade that may be obtained and hence there will
be less necessity for surface drainage, yet this part of
the work should never be lost sight of.

There are, however, valleys or depressions like large
ponds or swamps for which it is impracticable to maintain any surface drainage.   For such places we must

LAYING TWELVE-INCH DRAIN-TILE.

(*To face page* 146.)

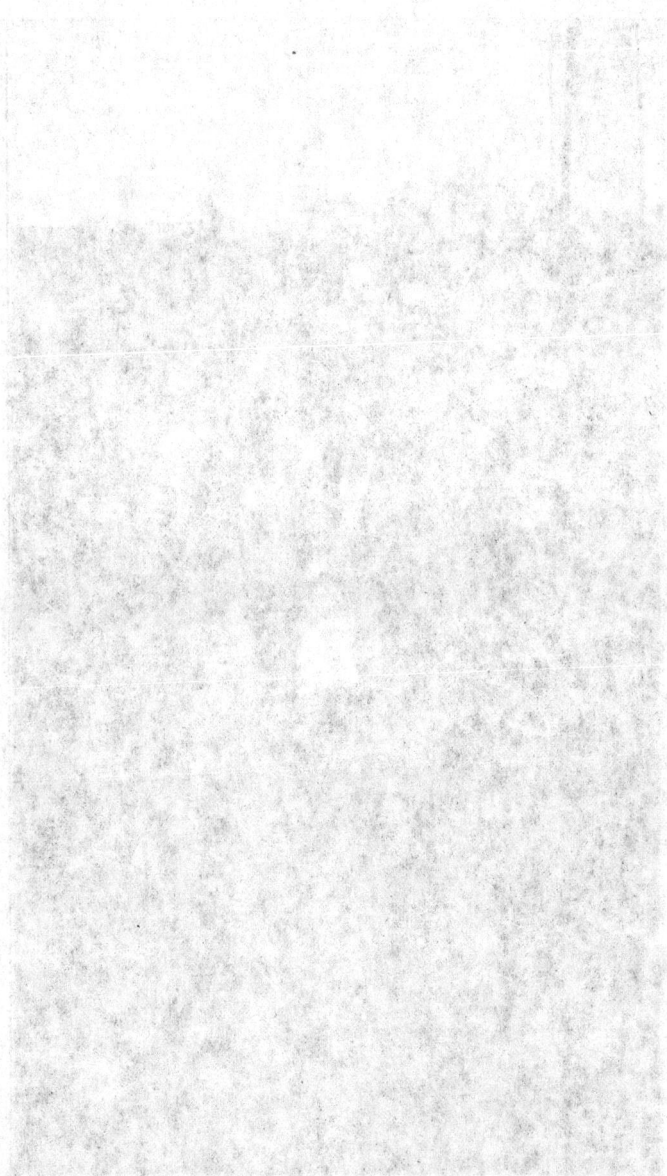

increase the size of the main tile, by figuring to carry $\frac{1}{2}$ inch of water in twenty-four hours, thus doubling the size that would otherwise be required.

In the consideration of questions relating to the size of drains, and provision for surface overflow, it will be seen that there is room for the exercise of knowledge and good judgment, and it may be added for the information of those who are sceptical regarding the value of the care thus far urged, that it is necessary if it is expected to realize the best results from the drainage work.

Selection of Tile.

The tile selected should be well burned, and hard enough to give a clear ring when struck. They should be uniform in size, so that each size can be laid in a smooth and continuous line without ragged projections on the inside. The approved form is the round tile 1 foot long. Sizes larger than 10-inch are better if made 24 inches long. There are many different qualities of tile, owing to the variety of clays that are worked for this purpose. The two general classes are known as "red tile" and "vitrified tile." The first are made of common clay and are very generally used for farm work. They are not always red in color, but are not vitrified and come under the general class. There are comparatively few of these that will endure alternate freezing and thawing such as the outlet of main drains is subject to, so that for such exposed places vitrified tile should be substituted. Many of them will not endure the exposure of the winter, when placed in piles on cultivated land, without scaling or chipping off.

When placed in the ground, however, they are durable and in every way satisfactory, provided they are well burned.   The vitrified tile are made of clay which will endure such heat that the peculiar elements which are a part of their composition will form a glassy material which is hard and as durable as the best quality of stone.   They are always desirable drain-tile.

The manner of designating the size is by naming the inside diameter, as 4-inch, 6-inch, meaning tile are 4 inches and 6 inches inside diameter respectively.   They are sold by the thousand feet or by the thousand pieces, each piece supposed to be 1 foot long.

The required number of junction tile for connecting the several laterals with the mains should be specified, as the practice of making them in the field by chipping and breaking the tile is not to be commended. The Y junction in preference to the T junction should be selected.   Curved tiles, such as are used in making curves in sewers, are not required for drains. The turns can be made sufficiently long so that the ordinary tile can be used if care is taken in fitting them. There is a tendency on the part of some manufacturers to make the walls of tiles too thin, which causes them to break too easily in handling.   The following is a safe minimum thickness for the walls of drain tiles:

3 inch....................	$\frac{1}{2}$ inch
4 to 5 "	$\frac{5}{8}$ "
6 and 7 "	$\frac{3}{4}$ "
9 "	$\frac{7}{8}$ "
10 and 12 "	$\frac{15}{16}$ "
14 "	1 "
15 "	$1\frac{1}{8}$ inches
18 "	$1\frac{1}{4}$ "

Tabulating Tile.

After the engineer has determined the number and size of the tile for each drain, he should note it upon the field-book along with other particulars pertaining to the drain. The tile of the whole field or system should then be tabulated, an example of which is given below, in order that a bill of tile according to size can conveniently be made out, and also that they may be distributed in the field without confusion. The following plan may be followed in making this list. The last column gives the total length of each separate drain and should be used in checking the work.

DISTRIBUTION OF TILE   (EXAMPLE OF FORM.)

Drain.	12-in	10-in	8-in	7-in.	6-in.	5-in	4-in	Total.
Main A..........	800	1200	250	450	200	1300	480	4680
No. 1.............							1350	1350
No. 2.............							1100	1100
No. 3							300	300
No 4					350	450	900	1700
Branch a of No. 4..						900	600	1500
No. 5.............						740	1260	2000
Main B.....			700		400	500		1600
No. 1 of B.......						200	400	600
No. 2 of B.......							600	600
	800	1200	950	450	950	4090	6990	15430

SUMMARY.

800. . . . . 12-inch

1200. . . . . 10 "

950. . . . . 8 "

450. . . . . 7 "

950. . . . . 6 "

4090. . . . . 5 "

6990. . . . . 4 "

_____

15430

CHAPTER XI.

OPEN DRAINS.

SUCCESSFUL tile drainage necessitates the construction of open ditches, unless the mains are so situated as to discharge into natural streams. When the area is too great to be drained by a tile main whose cost will be within profitable limits, then an open ditch should be used. It is sometimes difficult to determine when an open ditch should take the place of tile drain. Usually where the cost of the two is about equal the tile should be used provided the outlet for the tract will be just as efficient. However, this is a question that must be decided in the light of such facts and considerations as pertain to each individual case, and in accordance with sound theory and safe practice.

It is proposed to discuss this matter in such a practical way that what is said may be of use to the young engineer in laying out and constructing large open ditches for drainage purposes, not in a guess-work sort of a way, but according to the best knowledge and practice now existing.

Grade for Open Ditches.

The total fall across a tract has been fixed by nature, but it is the work of the engineer to determine whether or not nature has provided enough for his purpose, and to adapt the size, form, and cross-section of the ditch

to the available grade. The fall that is usually found
in districts where it is necessary to construct large outlet channels is from 1 foot to 5 feet per mile. These
are the main channels or outlets for the drainage of
large areas. Lateral open ditches for small areas and
for farms have usually a higher rate of fall, but not
always. It is desirable that ditches have sufficient fall
to be self-cleaning. This in soil and clay which is not
easily displaced is about 4 feet per mile, which for
ditches of ordinary size will give a mean velocity of
$2\frac{3}{4}$ miles per hour when running full. This is given
as an approximate grade for ditches which may be relied upon as self-cleaning, and applies to those with
bottoms not less than 3 feet wide and a depth not
less than 4 feet, and so adjusted in size that they will
run three fourths full at flood height. This is by no
means the minimum grade upon which effective ditches
may be constructed. As a matter of fact there are many
large tracts of land drained by outlet ditches having a
fall of from 6 to 24 inches per mile. These are not, however, self-cleaning except when made deep and large
with a final free discharge, and even then a liberal
annual allowance should be made for cleaning. As
can be readily understood, there can be but little or
no scouring action of the water upon the bottom of the
channel unless the head is furnished by the depth of
water in the ditch. Hence light-grade channels must
be deep—6 to 8 feet if possible—or else provision must
be made for frequent artificial cleaning. This continual
expense will in the course of ten years go far towards
paying the additional first cost of a deeper ditch.
Where tracts are level and grades light, depth of ditch

has come to be recognized a necessary adjunct of successful drainage.

Mean Velocity of Flow.

The velocity of water flowing in an open channel is retarded by the contact of the particles with the bottom and sides of the channel, these resistances being greater or less according to the nature of the material through which the channel is dug and the irregularities presented by the bottom and sides and having contact with the water. It has been demonstrated that the films of water from the bottom of the channel upward toward the surface and from the sides toward the center of the channel form vertical and horizontal curves respectively with the advanced portion of the curves in the center line of the stream. If these curves were platted the resistance of the sides and of the bottom of the ditch would have the appearance of holding back the water so that no two films of water would have the same velocity. The greatest velocity of a stream is found in the thread of the current just underneath the surface, all other parts having a less velocity in proportion as they approach the bottom and sides of the channel. In considering the discharge of a channel we must use the mean velocity of flow, which in a trapezoidal earth channel is about eight tenths of the surface velocity, and is the velocity found at a point in the center line of the stream and a little more than half way from the surface to the bottom. The bottom velocity is about seven tenths of the surface velocity, depending much upon the kind of material which forms the bottom. An increase in the depth of the water in a ditch

A DREDGED DRAINAGE-DITCH TEN YEARS AFTER EXCAVATION.

(To face page 192.)

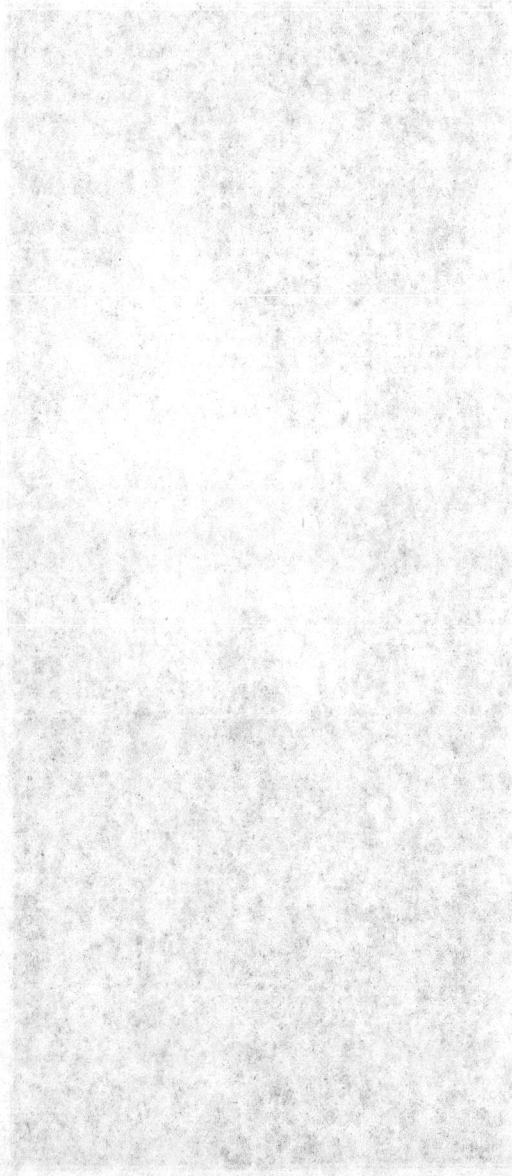

since it is a virtual addition of head, accelerates the velocity. The following table shows the effect that increase of depth has upon the mean velocity in a rectangular channel 10 feet wide with a grade of 3 feet per mile.

MEAN VELOCITY OF WATER AT DIFFERENT DEPTHS IN RECT-
ANGULAR DITCH 10 FT. WIDE, GRADE 3 FT. PER MILE.

Depth in Feet.	Mean Velocity in Feet per Second.
0.5	1.4
1.5	2.3
2.0	2.6
2.5	2.8
3.0	2.9
4.0	3.2
5.0	3.4
6.0	3.6
8.0	3.8

Relation of Breadth and Depth of Channel to Surface and Mean Velocity.

As the result of many experiments by careful hydraulicians, the following table from "Fanning's Hydraulic Engineering" may be given to show the relation of breadth, depth, surface velocity and mean velocity to each other for rectangular smooth channels when the water is from 5 to 10 feet deep. Let $b =$ breadth, $d =$ depth, $V =$ surface velocity at thread of current, and $v =$ mean velocity.

When $b = 2 d$ then $v = .920 V$.
" $b = 3 d$ " $v = .910 V$.
" $b = 4 d$ " $v = .896 V$.
" $b = 5 d$ " $v = .882 V$.
" $b = 6 d$ " $v = .864 V$.
" $b = 7 d$ " $v = .847 V$.
" $b = 8 d$ " $v = .826 V$.
" $b = 9 d$ " $v = .805 V$.
" $b = 10 d$ " $v = .780 V$.

The mean velocity will be a little less for a trapezoidal channel and will decrease as the side slopes
are flattened.   When the breadth is twice the depth
we have a mean velocity of flow equal to 92 per cent
of the surface velocity measured in the middle of the
channel.

Curvature of Ditches.

The proper curve to give ditches when they are deflected from a straight line is a matter which merits
careful attention.   It is desirable that the adjustment
of curve to velocity of flow be such that the banks will
not require artificial protection.   The relation of bank
erosion to curvature of the ditch and the velocity of
flow is intricate owing to the great difference in the
stability of earth when subjected to the action of water.

Circular curves are described by the number of degrees of arc which a chord of 100 feet subtends.   The
degree of a curve is determined by the central angle
which is subtended by a chord of 100 feet.   For curves
of from 1 degree to 10 degrees the radius may be found
by dividing 5730 feet (the radius of a 1-degree curve)
by the degree of the curve.   The following is a list of
curves and their corresponding radii, which may be
used as a basis in constructing ditches with limitation
as hereafter described:

Degree of Curve.	Radius in Feet.	Degree of Curve.	Radius in Feet.
7°	.819	14°	.410
8°	.716	15°	.383
9°	.637	16°	.359
10°	.574	17°	.338
11°	.522	18°	.320
12°	.478	19°	.310
13°	.442	20°	.288

While circular curves may be used to describe approximately the curvature that should be given, the
true form should not be geometrical, but rather what
may be termed natural, or such as is used in laying
out artificial streams and roads in parks, in which geometrical lines are ignored.   The difference between the
two is shown in Fig. 29, which is a 12-degree curve

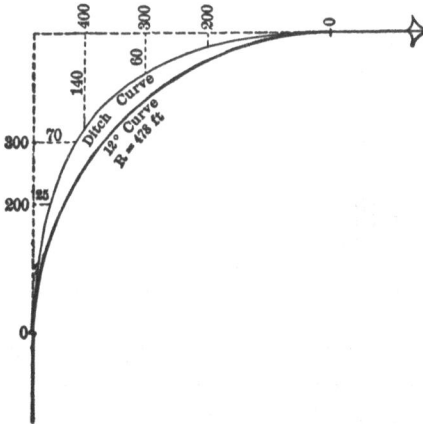

FIG. 29.—Proper Curve for Open Ditches.

(radius 478 feet) varied so as to subject the bank against
which the stream strikes when first deflected to the
least possible erosion.   The reason for this is well illustrated by Fig. 30, in which the stream is represented
as being divided into filaments, each having a velocity
imparted to it by the flow, and striking the opposite
bank as an individual force.   According to the well-
known law of force, the angles of incidence and reflec-
tion are equal when a force meets a resisting plane.
Hence in the case under consideration, the reflected

force is thrown against the other forces or filaments toward the interior of the stream and assist in breaking the force and deflecting the current. The section of curve first struck will receive the greatest force, and be subject to greater erosion if the curve were a segment

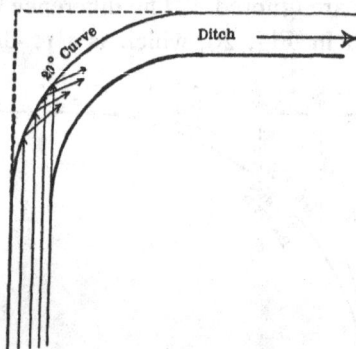

FIG. 30.—Action of Current on Ditch Banks at Curves.

of a circle. For this reason the up-stream part of the curve should be deflected from the tangent by using a curve of greater radius than the remainder of the curve, in order that all parts may be subject to uniform erosion.

When the points of tangency have been fixed upon, the curve may be "run in by the eye" better than by the instrument, and the center line located by measurements from the tangents in the manner shown in Fig. 29.

How short a curve may be used in large ditches such as are constructed for drainage districts without endangering the stability of the banks at the curve is a question that cannot be answered with mathematical certainty for the reasons previously stated. Deductions from close observations of both natural and artificial streams which flow through alluvial soils are the only guides to the work. From such observations the following empirical rules may be deduced:

For ditches with minimum bottom width of 4 feet and maximum grade of 2 feet per mile use 20-degree curve = radius of 288 feet. For ditches with bottom width 4 feet to 8 feet and grade of 3 feet to 6 feet per mile use 12-degree curve = 478 feet.

For larger ditches and greater fall, or for the above-named ditches which have a greater fall than indicated, curves ranging between 6 degrees and 12 degrees may be used with such latitude as conditions of earth and fall may indicate to the careful designer.

Form and Depth of Ditches.

The foregoing facts have an important bearing in determining the form that should be given to a ditch. A channel with vertical sides offers the least resistance to the current, so that if such a form could be maintained, it would carry a greater volume of water in proportion to its cross-sectional area than any other form. It will be also seen that the velocity is greater in a full channel than in one partly full, and in a deep channel running full than in a shallow one running full.

It follows, then, that the form of a ditch where the fall is light, and where it is desired to secure the best results with the least excavation, should approach as near as possible to that of a rectangle, and should be as deep as practicable. Nothing but rocky material will stand perpendicular. Ordinary clays will stand very well at a slope of 45 degrees, or 1 to 1 as it is called. Loose, loamy, and sandy soils will sometimes require a side slope of $1\frac{1}{2}$ to 1 in order to stand reason-

ably well.   The slope that is preferable in ordinary
soils is that of 1 to 1.   After a ditch made in this form
has been in use a year or two, the action of the water
will change its form, the upper part becoming more
nearly perpendicular and the lower more nearly horizontal.

In the laying out of ditches for the drainage of large
areas, a depth of from 5 to 7 feet should be aimed at
for two reasons.   First, this depth is needed for the
purpose of giving good lateral drainage, and second,
to get the velocity necessary to scour the bottoms and
make them self-cleaning as far as possible.   Broad flat
bottom shallow ditches are adapted to carry floods in
freshet times, but not to give outlets for the thorough
drainage of level tracts.   In this connection it may be
said that the bottom width of any ditch with a grade of
1 to 2 feet per mile should not be less than 3 feet.

Ditches with a grade of 10 feet per mile may be
made with any bottom width which will furnish the
necessary capacity.   Such ditches will be self-cleaning,
and often precautions must be taken against injury of
the ditch by erosion.   The side slopes must be governed by the kind of soil and clay through which they
are made, and also by the means that are to be used in
excavation.   If the work is to be done with teams, a
slope of about 2 to 1 must be given to the sides.   If
done with the steam dredge it can be made with slope
of 1 to 1 or less.

Capacity Required for Open Ditches.

How large a drainage-channel should be made to
afford an outlet for a given area is a question which

must be answered at the beginning of every drainage enterprise. There are three distinct elements that must be taken into consideration in deciding this matter.

1. *Area of the drainage-basin*, and also its shape; that is, the number of acres that will be drained by the ditch when made, and whether this area lies in a broad tract on either side of the proposed ditch, or is in the form of a narrow strip along its course.

2. *Slope of the land;* that is, whether it is broken up by alternate steep slopes and flats or is a plane having but little variation in level.

3. The fall that can be obtained for the ditch along the various parts of its course.

The bearing which these conditions have upon the size of ditches may be stated in a general way as follows: If the tract is level and broad, drainage-water will be held back a longer time and be distributed to the main ditch much more slowly than if the tract were narrow and has steep slopes which will shed water rapidly toward the main. If the surface is broken up and slopes are steep the soil will absorb less, and the greater surface slope will impel the surplus water with greater velocity over the surface toward the main ditch, and thereby tax its capacity more than if the land were more nearly level.

Quantity of Water per Acre to be Removed.

It is a common opinion that if a ditch of certain capacity will drain a given area that it will require a ditch of double that capacity to drain twice the area. An examination of natural streams and their water-sheds, as well as experience with artificial ditches, does not sustain this opinion. The larger the area the greater the proportion of drainage-water which is held back by obstructions, so that the discharge of the outlet ditch for a given time is not so great, neither is the proportion which is finally discharged as great for large as for small areas. The quantity of water to be carried per acre is greater for open ditches than for tile drains, since we have assumed that much of the flood water is to be carried by surface drains in the case of large tile-drained areas, and hence we must make provision for carrying the final surplus through the main channels which are now under consideration.

From a large number of observations upon areas which have been drained by ditches whose discharging capacities were known, it has been found that calculations should be made upon the following basis, in which it is assumed that at flood time the main ditch is to run eight tenths full.

Areas from 1000 to 3000 acres, with lateral fall of land toward the main at the rate of 2 to 3 feet per mile, remove $\frac{3}{4}$ inch in depth of water in twenty-four hours.

Areas 3000 to 8000 acres remove $\frac{1}{2}$ inch in depth in twenty-four hours.

Areas 8000 to 30,000 acres remove $\frac{1}{4}$ inch in depth in twenty-four hours.

If the land is rolling so that considerable areas have slopes of 1 to 3 or more feet per 100, the above figures should be increased proportionally. For ordinary drainage districts the $\frac{1}{2}$-inch standard will usually apply, understanding that very heavy rainfall will very nearly fill the ditch.

Formula for Open Ditches.

The force which impels water along the ditch is the rate of fall. The forces acting against this are the resistance offered to the water by the sides and bottom of the ditch. The difference between these forces is the velocity head and causes whatever current there may be produced. When the water becomes low in a ditch which has slight fall, the bottom being covered with silt or overgrown with grass, the two forces balance and there is no flow.

A formula used for velocity of flow in ordinary drainage-canals with good results is the following:

$$v = \sqrt{\frac{a}{p} \times 1\tfrac{1}{2} f.} \quad . \quad . \quad . \quad (10)$$
$$Q = av.$$

Where v = mean velocity in feet per second.

a = area of waterway.

p = wet perimeter or border of the channel which is wet.

f = fall in feet per mile.

Q = discharge in cubic feet per second.

This formula gives results sufficiently close for the purpose. It must be remembered that in applying any of the drainage formulas the data at hand are not always of the most accurate kind, and there must be a margin allowed for contingences that cannot be provided for by means of definite quantities which may be inserted in a formula.

To Apply the Formula.—Find the area of the water-way of the proposed ditch in square feet when it is

filled eight tenths full.   Find the length of channel
which is wet by this depth of water, called the wet
perimeter.   Find the fall in feet per mile and substitute the quantities thus determined in the formula and
find the value of v.

To find the number of acres which the proposed
ditch will drain, proceed as directed for finding the
number of acres which a tile will drain.   Find the
value of Q and divide it by the decimal which represents
the quantity of water that must be discharged per second in order to relieve the land of the desired depth of
water per acre.   By formula the expression would be

$$A = \frac{Q}{n}. \quad \cdots \quad \cdots \quad \text{(11)}$$

A = number of acres, n = number taken from table
No. 5.

Example: Given a ditch with bottom 8 feet, fall 3
feet per mile, side slopes 1 to 1, water to run 5 feet
deep in flood time, how many acres will it give an
outlet for using the $\frac{1}{2}$-inch standard?

$a = 13 \times 5 = 65.$

$p = 8 + 2 \sqrt{25 + 25} = 22.14.$

$1\frac{1}{2} f = 4.5$ Substituting in the formula

$v = \sqrt{\dfrac{65}{22.14}} \times 4.5 = 3.63,$

$Q = 65 \times 3.63 = 235.9 = \text{cu. ft. disch. per sec.},$

$A = \dfrac{235.9}{.0210} = 11,233.$

With a fall of 2 feet per mile $A = 9183.$
"   "   "   "   " 1 foot "   " $A = 6914.$

The following examples of ditches of different sections and rates of fall, with computations for velocity, discharge, and number of acres which they will drain (Fig. 31), will serve to show the method of using

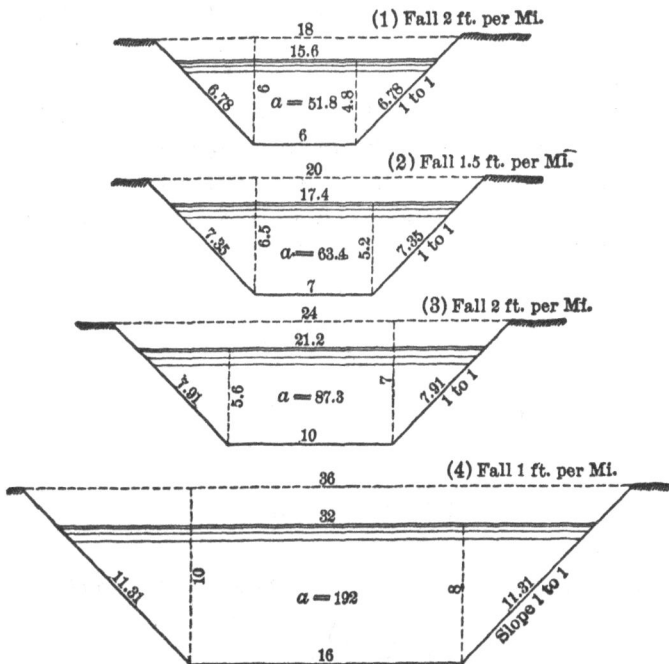

FIG. 31.—Data required for Computing Discharge and Acreage Drained by Ditches.

formula (10). This method of determining the number of acres for which a ditch will give an outlet is given because it is based on correct principles, is simple of application, and has been found applicable to level tracts of land. Its adaptability to lands of different slopes with a variety of contour lies in the

use of the different standards for the quantity of water
that it is desired to remove. For example, if the tract
through which the ditch passes is rolling, so that the
lateral drainage is brought to the main ditch more
rapidly than is indicated by the fall of the main, there
is a greater run-off in a given time, and hence the
higher rate of discharge must be used in computing the
number of acres that a ditch will serve. A more elaborate hydraulic formula more correct theoretically
would require a greater refinement of data than can
usually be obtained for such work within reasonable
cost limits

(1)

$$v = \sqrt{\frac{51.81}{19.56}} \times 3 = 2.81,$$

$$Q = 145.6, \qquad\qquad A = \frac{145.6}{.021} = 6931.$$

(2)

$$v = \sqrt{\frac{63.44}{21.70}} \times 2.25 = 2.56,$$

$$Q = 162.4, \qquad\qquad A = \frac{162.4}{.021} = 7733.$$

(3)

$$v = \sqrt{\frac{87.36}{25.82}} \times 3 = 3.18,$$

$$Q = 277.8, \qquad\qquad A = \frac{277.8}{.021} = 13223.$$

(4)

$$v = \sqrt{\frac{192}{38.62}} \times 1.5 = 2.73,$$

$$Q = 524, \qquad\qquad A = \frac{524}{.021} = 25000.$$

Waste Banks and Berm.

The waste banks from the large ditches often become a serious inconvenience in the cultivation of fields, and where the work is done by the steam-dredge their reduction to suitable proportions for farming purposes is a matter which will require time as well as labor. The earth is deposited in a wet and plastic condition, and if it contains no sand or gravel becomes exceedingly hard and tough when dry or partially dry. It resists all efforts that may be made to spread or smooth the irregular heaps. Nothing but exposure to the weather will so disintegrate the mass that it can be graded or scraped back from the ditch. A due regard for economy will dictate that the work of reducing the waste banks shall be done gradually, as the action of sun and rain and frost may open and pulverize the surface. The waste banks also form a dike or barrier on each side of the ditch, which prevents the water of drainage depressions along the course from entering the ditch. Breaks or openings should be left during construction at those depressions which may subsequently be required to perfect lateral surface drainage. It is always preferable to keep a strip of land 10 feet wide on each side of the ditch seeded to grass, as this will secure the banks by reason of the turf and also prevent the surface run-off along the edge of the ditch from carrying loose earth into it.

The width of berm or clear space between the edge of the ditch and the waste bank is important. The weight of the excavated earth should be so far away that the sides of the ditch when saturated with water will not give way and slide by reason of the superincumbent weight. A distance of 7 feet from the outer edge of the ditch to the inner edge of the waste bank will be sufficient, yet more than this will make the after working down of the waste bank more convenient.

Surveys for Open Ditches.

After having done the necessary preliminary topographical work for the purpose of determining in a general way where the ditch or ditches should be located, the lines should be surveyed, the grades established, and the quantities of earth to be excavated computed. The first work is to run the center line of the ditch, which is done in the manner described for running a line for a tile drain. Stakes should be set, levels taken, and the line located for mapping purposes as heretofore described.

Locating the Grade.—The most satisfactory method of determining upon a suitable grade for a long line is to first make a profile upon profile paper, which will show to the eye the relation of different points along the line with respect to their elevations. Draw trial grade lines upon this profile in pencil until one is found which will give a desired fall and depth to the several sections of the ditch. Then run this in on the notes and compute the cuts as directed for tile drains. If the line is short, say a mile long, the grade may easily be fixed in the manner described for tile drains. In considering the matter of grade and depth, it should always be found whether or not all of the land that it is desired to drain can be well drained through the proposed ditch, and if not, whether the grade and depth can be so adjusted as to furnish the desired outlet. These points can be determined by comparing the elevations furnished by the topographical survey if one has been made. If such a survey has not been made, levels should be run out from the main line to all doubtful points. At the time the survey of the center line is made, "bench-marks" should be fixed at convenient places not far from the line, to which reference can be made in testing the ditch during and after construction. The most permanent are made on the brace roots of trees into which a notch is cut. Usually, however, solid hub stakes driven firmly into the ground at out-of-way points must be used. They should be driven flush with the surface and described in the notes after being marked by some guide-post or stake, by which they may be located.

Cross-sectioning.—Having established a grade, figured the center cut, and decided upon the size and form of the ditch, proceed to set the slope stakes which will define the top width of the ditch and to take such levels as may be necessary to properly compute the required excavation.

Where ditches are made through level tracts and no deductions are to be made for ditches already existing, the distance out from the center may be measured direct and a level taken at the side for a guide in construction. In case the ground is more or less uneven, the setting of slope stakes is a necessary part of the work for both construction and computation purposes.

The following is a convenient form for taking and keeping all of the necessary notes. The center line is

supposed to have been run first and computed and the
necessary blank lines left for the subsequent cross-
section work.

FORM OF NOTE-BOOK FOR OPEN-DITCH SURVEYS.

Sta.	+ S.	H. I.	− S.	Elev.	G. L.	C. Cut.	R. Cut.	L. Cut.	Distance Out.		Cu. Yds.
									R.	L	
0.......	...	...	...	...	...	...	...	...	...	...	...
Right...	...	...	...	...	...	...	...	...	...	...	...
Left	...	...	...	...	...	...	...	...	...	...	...
I−	...	...	...	...	...	...	...	...	...	...	...

Knowing grade, center cut, bottom width, and side
slopes, set up the instrument and obtain the H. I. from
center hub or from the nearest bench-mark. If the
ground is level each side of the center, the distance out
from the center for slope of 1 to 1 will be $\frac{1}{2}$ bottom
width plus center cut, for slope of 1$\frac{1}{2}$ to 1 will be $\frac{1}{2}$
bottom width plus 1$\frac{1}{2}$ times center cut, etc.   For
ground higher or lower than the center, take a level
at an estimated distance, find elevation, subtract from
it the grade-line elevation and obtain cut.   Use this
cut instead of the center cut as above described, and if
the computed distance out corresponds with that measured, drive the slope stake and record the cut and
distance out in their proper place in the book.   If not,
move the rodman out or in until the proper point has
been found.   The rodman measures the distance out
with his rod and drives the stakes as directed.   It is
good practice to set a side hub on the right-hand side
for the use of the contractor while constructing the
ditch, since the centers must all be destroyed, but a side

stake can be preserved for reference at any time. The center and side cuts and distance out when recorded in the notes furnish the data required for computing the excavation. If the ground is very irregular, a sufficient number of sections must be taken to enable the engineer to compute the contents within reasonable limits of accuracy.

The slope stakes having been set, the contractor may begin excavation at the limit indicated by them and carry the required slope to the required depth, at which depth the ditch will have the width designated.

Computing the Cubic Yards of Excavation.—In the usual class of work required for drainage canals it is not necessary to use the lengthy and inconvenient method of computing earthwork by the prismoidal formula. However, in very rough ground it should be used. This may be stated as follows: From the notes compute the end areas of a 100-foot section or station, from these construct the middle area. Add to four times the middle area the area of end sections and take one sixth of the product for the mean area. Fractional parts of a station should be treated in the same manner if the field notes have been taken for that purpose. The method by end areas is as follows: Add the end areas of any given station and divide by two and the result is the mean area. This in either of the above methods when multiplied by the length of station and divided by 27 will give the number of cubic yards in the station. The work may be very much shortened by the use of the following table:

TABLE 7.
EXCAVATION TABLES.

Area in Feet	0.00	0.10	0.20	0.30	0.40	0.50	0.60	0.70	0.80	0.90
0	0.00	0.37	0.74	1.11	1.48	1.85	2.22	2.59	2.96	3 33
1	3.70	4.07	4.45	4.81	5.19	5.56	5.93	6.30	6.67	7.04
2	7.41	7.78	8.15	8.52	8.89	9.26	9.63	10.00	10.37	10 74
3	11.11	11.48	11.85	12.22	12.59	12.96	13.33	13.70	14.07	14.44
4	14.82	15.19	15.56	15.93	16 30	16.67	17.04	17.41	17.78	18 15
5	18.52	18.89	19.26	19.63	20.00	20.37	20.74	21.11	21.48	21.85
6	22.22	22.59	22.96	23.33	23.70	24.07	24.44	24.82	25.19	25.56
7	25.93	26.30	26.67	27.04	27.41	27.78	28.15	28.52	28.89	29.26
8	29.63	30.00	30.37	30.74	31.11	31.48	31.85	32.22	32.59	32.96
9	33.33	33.70	34.07	34.44	34.82	35.19	35.56	35 93	36.30	36.67
10	37.04	37.41	37.78	38.15	38.52	38.89	39.26	39.63	40.00	40.37
11	40.74	41.11	41.48	41.85	42.22	42.59	42.96	43.33	43.70	44.07
12	44.44	44.82	45.19	45.56	45.93	46.30	46.67	47.04	47.41	47.78
13	48.15	48.52	48.89	49.26	49.63	50.00	50.37	50.74	51.11	51.48
14	51 85	52.22	52.50	52.96	53.33	53.70	54.07	54.44	54.82	55.19
15	55.56	55.93	56.30	56.67	57.04	57.41	57.78	58.15	58.52	58.89
16	59.26	59.63	60.00	60.37	60.74	61.11	61.48	61.85	62.22	62.59
17	62.96	63.33	63.70	64.07	64.44	64.82	65.19	65.56	65.93	66.30
18	66.67	67.04	67.41	67.78	68.15	68.52	68.89	69 26	69.63	70.00
19	70.37	70.74	71.11	71.48	71.85	72.22	72.50	72.96	73.33	73.70
20	74.07	74.44	74.82	75.19	75.56	75.93	76.30	76.67	77.04	77.41
21	77.78	78.15	78.52	78.89	79.26	79.63	80 00	80.37	80.74	81.11
22	81.48	81.85	82.22	82.59	82.96	83.33	83.70	84.07	84.44	84.82
23	85.19	85.56	85.93	86.30	86.67	87.04	87.41	87.78	88.15	88.52
24	88.89	89.26	89.63	90.00	90.37	90.74	91.11	91.48	91.85	92.22
25	92.59	92.96	93.33	93.70	94.07	94.44	94.82	95.19	95.56	95.93
26	96.30	96.67	97.04	97.41	97.78	98.15	98.52	98.89	99.26	99.63
27	100.00	100.37	100.74	101.11	101.48	101.85	102.22	102.59	102.96	103.33
28	103.70	104.07	104.44	104.82	105.19	105.56	105.93	106.30	106.67	107.04
29	107.41	107.78	108.15	108.52	108.89	109.26	109.63	110.00	110.37	110.74
30	111.11	111.48	111.85	112.22	112.59	112.96	113.33	113.70	114.07	114.44
31	114.81	115.18	115.56	115.92	116.29	116.67	117.03	117.40	117.77	118.15
32	118.52	118.89	119.26	119.63	120.00	120.37	120.74	121.11	121.48	121.85
33	122.22	122.59	122.96	123.33	123.70	124.07	124.44	124.81	125.18	125.55
34	125.92	126.30	126.66	127.03	127.40	127.77	128.14	128.51	128.88	129.26
35	129.63	130.00	130.37	130 74	131.11	131.48	131.85	132.22	132.59	132.96
36	133.33	133.70	134.07	134.44	134.81	135.18	135.55	135.92	136.29	136 67
37	137.04	137.41	137.78	138.15	138.52	138.89	139.26	139.63	140.00	140.37
38	140.74	141.11	141.48	141.85	142.22	142.59	142.96	143.33	143.70	144.07
39	144.44	144.81	145.18	145.55	145.92	146.29	146.66	147.03	147.40	147.78
40	148.15	148.52	148.89	149.26	149.63	150.00	150.37	150.74	151.11	151.48
41	151.85	152.22	152.59	152.96	153.33	153.70	154.07	154.44	154.81	155.18
42	155.55	155.92	156.29	156.66	157.03	157.40	157.77	158.14	158.51	158.89
43	159.26	159.63	160.00	160.37	160.74	161.11	161.48	161.85	162.22	162 59
44	162.96	163.33	163.70	164 07	164.44	164 81	165.18	165.55	165.92	166.30
45	166.67	167.04	167.41	167 78	168.15	168.52	168.89	169.26	169.63	170.00
46	170.37	170.74	171.11	171 48	171.85	172.22	172.59	172.96	173.33	173.70
47	174.07	174.44	174.81	175.18	175 55	175.92	176.29	176.66	177.03	177.41
48	177.78	178.15	178 52	178.89	179.26	179.63	180.00	180.37	180.74	181.11
49	181.48	181.85	182.22	182 59	182.96	183.33	183.70	184.07	184.44	184.81
50	185 18	185.55	185.92	186.29	186.66	187.03	187.40	187.77	188.14	188.52
51	188.89	189.26	189.63	190 00	190.37	190.74	191.11	191.48	191.85	192.22
52	192.59	192.96	193.33	193.70	194.07	194.44	194.81	195.18	195.55	195 93
53	196.30	196.67	197.04	197 41	197 78	198.15	198.52	198.89	199.26	199 63
54	200.00	200 37	200.74	201.11	201 48	201.85	202.22	202.59	202.96	203.33
55	203.70	204.07	204.44	204 81	205 18	205.55	205.92	206.29	206.66	207.03
56	207.41	207.78	208.15	208.52	208.89	209.26	209.63	210.00	210.37	210 74
57	211.11	211.48	211 85	212.22	212.59	212.96	213.33	213.70	214.07	214.44
58	214.81	215.18	215.55	215.92	216.29	216.66	217.03	217.40	217.77	218.15

TABLE 7—*Continued.*

Area in Feet	0.00	0.10	0.20	0.30	0.40	0.50	0.60	0.70	0.80	0.90
59	218.52	218.89	219.26	219.63	220.00	220.37	220.74	221.11	221.48	221.85
60	222.22	222.59	222.96	223.33	223.70	224.07	224.44	224.81	225.18	225.55
61	225.92	226.29	226.66	227.03	227.40	227.77	228.14	228.51	228.88	229 26
62	229.63	230.00	230.37	230.74	231.11	231.48	231.85	232.22	232.59	232.96
63	233.33	233.70	234.07	234.44	234.81	235.18	235.55	235.92	236.29	236.67
64	237.04	237.41	237.78	238.15	238.52	238.89	239.26	239.63	240.00	240.37
65	240.74	241.11	241.48	241.85	242.22	242.59	242.96	243.33	243.70	244.07
66	244.44	244.81	245.18	245.55	245.92	246.30	246.67	247.04	247.41	247.78
67	248.15	248.52	248.89	249.26	249.63	250.00	250.37	250.74	251 11	251.48
68	251.85	252.22	252.59	252.96	253.33	253.70	254.07	254.44	254.81	255.18
69	255.50	255.93	256 30	256.67	257.04	257.41	257.78	258.15	258.52	258.89
70	259.26	259.63	260.00	260.37	260.74	261.11	261.48	261.85	262.22	262.59
71	262.96	263.33	263.70	264 07	264.44	264.81	265.18	265 55	265.92	266.30
72	266.67	267.04	267.41	267.78	268.15	268.52	268.80	269.26	269.63	270.00
73	270.37	270.74	271.11	271.48	271.85	272.22	272.59	272.96	273.33	273.70
74	274.07	274.44	274.81	275.18	275.55	275.92	276.20	276.66	277.04	277.41
75	277.78	278.15	278.52	278.89	279.26	279.63	280.00	280.37	280.74	281.11
76	281.48	281.85	282.22	282.59	282.96	283.33	283.70	284.07	284.44	284.81
77	285.18	285.56	285.93	286.30	286.67	287 04	287.41	287.78	288.15	288.52
78	288 89	289.26	289.63	290.00	290.37	290.74	291.11	291.48	291.85	292.22
79	292.59	292.9	293.33	293.70	294.07	294.44	294.81	295.18	295.55	295.93
80	296.30	296.67	297.04	297.41	297.78	298.15	298.52	298.89	299.26	299.63
81	300.00	300.37	300.74	301.11	301.48	301.85	302.22	302.59	302.96	303.33
82	303.70	304.07	304.44	304.81	305.18	305.55	305.92	306.20	306.66	307.03
83	307.41	307.78	308.15	308.52	308.89	309.26	309.63	310.00	310.37	310.74
84	311.11	311.48	311.85	312.22	312.59	312.96	313.33	313.70	314.07	314.44
85	314.81	315.19	315.56	315.93	316.30	316.67	317.04	317.41	317.78	318.15
86	318.52	318.89	319.26	319.63	320.00	320.37	320.74	321.11	321.48	321.85
87	322.22	322.59	322.96	323.33	323.70	324.07	324.44	324.81	325.18	325.55
88	325.92	326.30	326.67	327.04	327.41	327.78	328.15	328.52	328.89	329.26
89	329.63	330.00	330.37	30.74	331.11	331.48	331.85	332.22	332.59	332.96
90	333.33	333.70	334.07	334.44	334.81	335.18	335.55	335.92	336.29	336.67
91	337.04	337.41	337.78	338.15	338.52	338.89	339.25	339.62	339.99	340.37
92	340.74	341.11	341.48	341.85	342.22	342.59	342.96	343.33	343.70	344.07
93	344.44	344.81	345.18	345.55	345.93	346.30	346.67	347.03	347.40	347.78
94	348.15	348.52	348.89	349.26	349.63	350.00	350.37	350.74	351.11	351.48
95	351.85	352.22	352.59	352.96	353.33	353.70	354.07	354 44	354.81	355.18
96	355.55	355.93	356.30	356.67	357.04	357.41	357.78	358.15	358.52	358.89
97	359.26	359.63	360.00	360.37	360.74	361.11	361.48	361.85	362.22	362.59
98	362.96	363.33	363.70	364.07	364.44	364.81	365.18	365.55	365.92	366.30
99	366.67	367.04	367.41	367.78	368.15	368.52	368.89	369.26	369.63	370.00
100	370.37	370.74	371.11	371.48	371.85	372.22	372.59	372.96	373.33	373.70
101	374.07	374.44	374.81	375.18	375.55	375.92	376.20	376.67	377.04	377.41
102	377.78	378.15	378.52	378.89	379.26	379.63	380.00	380.37	380.74	381.11
103	381.48	381.85	382.22	382.59	382.96	383.33	383.70	384.07	384.44	384.81
104	385.18	385.55	385.92	386.29	386.67	387.04	387.41	387.78	388.15	388.52
105	388.89	389.26	389.63	390.00	390.37	390.74	391.11	391.48	391.85	392.22
106	392.59	392.96	393.33	393.70	394.07	394.44	394.81	395.18	395.55	395.92
107	396.30	396.67	397.04	397.41	397.78	398.15	398.52	398.89	399.26	399.63
108	400.00	400.37	400.74	401.11	401.48	401.85	402.22	402.59	402.96	403.33
109	403.70	404.07	404.44	404.81	405.18	405.55	405.92	406.29	406.67	407.04
110	407.41	407.78	408.15	408.52	408.89	409.26	409.63	410.00	410.37	410.74
111	411.11	411.48	411.85	412.22	412.59	412.96	413.33	413.70	414.07	414.44
112	414.81	415.18	415.55	415.92	416.29	416.67	417.04	417.41	417.78	418.15
113	418.52	418.89	419.26	419.63	420.00	420.37	420.74	421.11	421.48	421.85
114	422.22	422.59	422.96	423.33	423.70	424.07	424.44	424.81	425.18	425.55
115	425.93	426.30	426.67	427.04	427.41	427.78	428.15	428.52	428.89	429.26
116	429.63	430.00	430.37	430.74	431.11	431.48	431.85	432.22	432.59	432.96
117	433.33	433.70	434.07	434.44	434.81	435.18	435.55	435.92	436.29	436.67
118	437.04	437.41	437.78	438.15	438.52	438.89	439.26	439.63	440.00	440.37
119	440.74	441.11	441.48	441.85	442.22	442.59	442.96	443.33	443.70	444.07

TABLE 7—*Continued.*

Area in Feet	0.00	0.10	0.20	0.30	0.40	0.50	0.60	0.70	0.80	0.90
120	444.44	444.81	445.18	445.55	445.92	446.29	446.67	447.04	447.41	447.78
121	448.15	448.52	448.89	449.26	449.63	450.00	450.37	450.74	451.11	451.48
122	451.85	452.22	452.59	452.96	453.33	453.70	454.07	454.44	454.81	455.18
123	455.55	455.92	456.29	456.67	457.04	457.41	457.78	458.15	458.52	458.89
124	459.26	459.63	460.00	460.37	460.74	461.11	461.48	461.85	462.22	462.59
125	462.96	463.33	463.70	464.07	464.44	464.81	465.18	465.55	465.93	466.30
126	466.67	467.04	467.41	467.78	468.15	468.52	468.89	469.26	469.63	470.00
127	470.37	470.74	471.11	471.48	471.85	472.22	472.59	472.96	473.33	473.70
128	474.07	474.44	474.81	475.18	475.56	475.93	476.30	476.67	477.04	477.41
129	477.78	478.15	478.52	478.89	479.26	479.63	480.00	480.37	480.74	481.11
130	481.48	481.85	482.22	482.59	482.96	483.33	483.70	484.07	484.44	484.81
131	485.18	485.55	485.92	486.29	486.67	487.04	487.41	487.78	488.15	488.52
132	488.89	489.26	489.63	490.00	490.37	490.74	491.11	491.48	491.85	492.22
133	492.59	492.96	493.33	493.70	494.07	494.44	494.81	495.10	495.56	495.93
134	496.30	496.67	497.04	497.41	497.78	498.15	498.52	498.89	499.26	499.63
135	500.00	500.37	500.74	501.11	501.48	501.85	502.22	502.59	502.96	503.33
136	503.70	504.07	504.44	504.81	505.18	505.56	505.93	506.30	506.67	507.04
137	507.41	507.78	508.15	508.52	508.89	509.26	509.63	510.00	510.37	510.74
138	511.11	511.48	511.85	512.22	512.59	512.96	513.33	513.70	514.07	514.44
139	514.81	515.18	515.55	515.92	516.29	516.67	517.04	517.41	517.78	518.15
140	518.52	518.89	519.26	519.63	520.00	520.37	520.74	521.11	521.48	521.85

To find the number of cubic yards from Table 7, turn to the left-hand column and find the corresponding area number. Opposite this will be found the number of cubic yards in a length of 100 feet. If the area has a decimal part pass the eye to the right and take the number of yards in the column under the decimal corresponding to the one required. If the number of yards for a part of a station only is required take such a part of the tabular number given as the required length is of 100 feet.

Examples: The mean area for a 100-foot section is 46. How many cubic yards of excavation? Find 46 in the left-hand column, and opposite is 170.37, the number of cubic yards.

Suppose the mean area of a 100-foot length is 60.7, find 60 in the left-hand column. Pass the eye to the right and in the column headed .70 take 224.81, the number of cubic yards in the station. As the several

stations are computed, enter the results in the note-
book opposite the respective stations, and in the column
headed for that purpose.

Ditching with Steam Dredges.

As large ditches can be made more profitably with
the steam dredge than by any other means, a brief description of the working of these machines is here given.

There are two general types of dredges used for this
work, differing mainly in the way they are moved over
the line of the work. One is the float dredge, in which
the machinery is mounted on a float boat, the engine
in the rear and the turn-table and excavating machinery
at the front. The excavator is in the form of a large
dipper holding from $\frac{1}{2}$ yard to $1\frac{3}{4}$ yards, according
to the size of the machine, and is operated like an
ordinary river dredge. The dipper is lowered to the
bottom of the ditch in front of the boat, filled, then
raised and swung to one side, and the contents dropped
by tripping a latch which allows the bottom of the dipper to fall. The boat is prevented from tipping over on
its side when the loaded dipper is swung by strong
braces with feet, which reach to the bottom or rest on
the bank of the ditch. As the excavation proceeds,
the boat is made to float forward as fast as desired.
The excavation is done under water, the depth of the
ditch being gauged mainly by the height of the water
on the dipper handle as it descends into the water to
be loaded. A necessary attendant of the dredge is
the floating boarding-house which follows a few hundred
feet in the rear. This is fitted up with kitchen,

dining-room, and sleeping-berths for the accommodation of the working force.

Another style is the drag-boat dredge, which begins operations at the outlet of the ditch and completes the work as it goes up-stream. The excavating machinery is similar to that just described, but it is mounted compactly on a boat about 7 feet deep, the sides having an angle of forty-five degrees and a bottom as wide as may be desired. Some are made as narrow as 4 feet on the bottom. The machine is moved forward by means of a wire cable which is anchored a few hundred feet ahead of the boat and attached to a winding drum placed underneath the deck of the boat. When it is desired to move the boat forward, the drum is set in motion, which winds up the cable and moves the boat forward. Little or no water is required for the successful operation of a machine of this kind. The ditch can be dug with any desired side slope, and the grade can be followed accurately by means of guides set ahead of the work in accordance with the survey. It will be noticed that the float dredge must have water to work in and begins at the upper end of the ditch and proceeds down the stream. The drag boat requires no water except to supply the boiler, begins at the outlet and propels itself up-stream. The first is adapted to large ditches where the water is plenty. The second is used most successfully for smaller ditches, where the water supply is small during the operating season.

The excavated earth is dropped on either side of the ditch, leaving a clear berm of from 6 to 8 feet as desired. The efficiency of these machines for doing this class of work is unquestioned. By means of them, large and deep ditches such as are required to drain from 2000 to 40,000 acres of level land can be profitably excavated.

The machines are above described to indicate the two general methods of working which are successfully used rather than to particularize machines.

Ditching with Teams.

Ditching with teams can be done profitably only under favorable conditions of earth and weather. Teams can be worked only where the animals can obtain a secure footing. The excavation is done with steel scrapers or "slips" after the earth has been loosened with the plough. Clay that contains a small per cent of sand will cut and dump with ease, while heavy clays of all kinds are exceedingly tough when found in their natural places in lands requiring ditches. However, the soil and subsoils are so various in different localities that nothing more can be said than that where this method is used every advantage of soil, season, and weather must be taken or the work may be blocked and laid over until the next favorable time. Road-graders with carriers for depositing the earth clear of the ditch-banks may be used where the earth will plough well and there is abundant room for operating. They require twelve horses and three men and will make a ditch about 2 feet deep with side slopes of 2 to 1 or more, except for large canals, where they can be used to advantage for any depth and width.

With reference to laying out and testing ditches which are made by teams it should be observed that every survey stake near the line of work will be destroyed during construction. The top width of the ditch should be laid out and defined by the stakes. The first work of the contractor should be to mark the outside lines of the ditch by a well-defined furrow, and he should also make a note of the depth of the proposed ditch at such points as he may be able to identify. The ditch should be inspected and tested for grade as the construction proceeds. To do this, begin at the zero or beginning point of the ditch, and by measurement reproduce the station points, take levels at these points on the bottom of the ditch, and compare the elevations obtained with those required. It will be remembered that bench-marks have been established at convenient intervals along the line for the purpose of reference during construction, and no ditch should be considered finished which has not been tested in this way.

Discharge of Side Ditches into Main.

Much trouble is experienced in alluvial soils or those which wash easily at points where lateral surface water enters the main ditch by reason of the earth which is washed into the main and is frequently found deposited in bars at or near the junction. This is particularly observed at points where a large main ditch crosses a public highway and the water from the shallow road ditches is discharged into the deeper main by an overfall which rapidly cuts away the earth in time of freshet. Much of this difficulty may be avoided by cutting the lateral ditches down to such a grade that the point of discharge will be near the bottom of the main as shown in Fig. 32. Provision should be made at all points where water discharges into open ditches in large quantities to cut off the overfall. Much injury to ditches and after expense in cleaning out may in this way be avoided. The line AB in the figure shows how the grade of a shallow ditch which discharges into a deeper one should be changed in order to avoid the

FIG. 32.—Section showing Junction of Shallow Ditch with Deep Main.

erosion of earth and consequent filling of the receiving ditch which will result if the side ditch is permitted to discharge on its regular grade by an overfall.

Special Forms for Ditches.

In practice, ditches are not made in the same form, but are adapted to the material through which they are made, the means used for their excavation, and the office they are to perform.

The cuts in Fig. 33 show forms of ditches which may be found in use and serve their purposes well. Fig. A is the common form left by the floating dredge in level districts where the depth is from 6 to 8 feet. B is a form used for a deeper ditch where clay through which it is cut is sufficiently firm to stand. C is a form used to provide for flood water from hills. For ordinary drainage the smaller channel is sufficient. This is adapted to a ditch with light grade, but which receives water from adjoining territory which has a heavy grade and which requires large flood capacity for a short time only.

D is the form of shallow scraper ditch suitable for surface drainage and overflow ditches. They may be

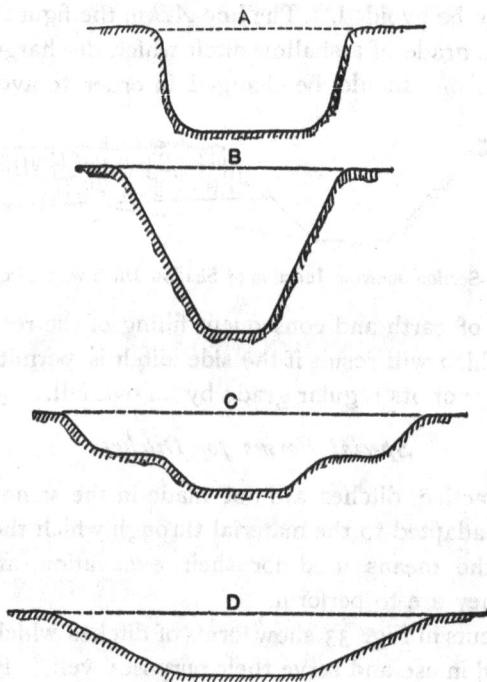

FIG. 33.—Sections of Open Ditches.

mere depressions which can be easily crossed with teams or may be moderately deep ditches with flat sloping sides which can be cleared of grass and weeds by the use of the mowing-machine.

Such a form of ditch should be selected as will be suited to the purpose, taking locality and work to be accomplished into account.

CHAPTER XII.

DRAINAGE OF BARN-YARDS, CATTLE-LANES, ETC.

THE stockman is familiar with the difficulties to be met in keeping ground which becomes puddled by the frequent tramping of live stock from becoming excessively muddy. Such ground cannot be materially benefited by placing tile drains underneath the puddled surface. The remedy consists in preventing all water from outside sources from finding its way to the yard, leaving only the direct rainfall to be contended with.

The roof water from all buildings adjoining the yards should be taken care of by eaves-troughs, and down-spouts which should conduct the water either into cisterns or into a tile drain provided for the purpose. This receiving drain should be laid around the buildings and discharge into some open channel or into a system of field drains. A tile of 8-inch diameter will carry the roof water from a large barn, and may discharge into a field main without overcharging it, for the reason that in all heavy rainfalls the roof water will have passed through the drain before soil water will have had time to enter. It is not desirable that stock-yards should be kept dry by surface drainage unless the value of the manure is to be disregarded. It is the practice of many good farmers to so arrange the stock-yards that all rainfall will gravitate toward the center, which thereby

becomes the receptacle for valuable manures and gives
drainage to the outer parts of the yard.   In all cases
the surface water from surrounding land should be cut
off by shallow trenches supplemented by underdrains.
These suggestions, if followed, will result in a great
amelioration of the mud evil so often endured by farmers and stockmen under the impression that there is
no remedy for the knee-deep conditions of their yards.

Taking Surface Water into Underdrains.

No surface water should be permitted to enter tile
drains direct unless precautions are taken to prevent
mud and débris from entering the drain.   It is frequently desirable to remove surface water by means of
underdrains from certain places which are not susceptible to drainage by soil filtration nor provided with
surface-drain outlets.   Some of these are ponds or depressions by the roadside, yards which are kept closely
compacted by constant use, drainage from carriage
washes and barns which is charged with muddy material, and other like necessities.

For such purposes the catch-basin shown in Fig. 34
will serve an excellent purpose.   It is a well constructed with brick, 3½ feet in diameter, with a depth
of 2 feet below the outlet pipe for the settling of mud
and heavy material which should be removed whenever the space below the discharge is filled.   The
discharge pipe should be not less than 6 inches in diameter in any case and should not be connected with
any extensive field system, but should extend direct
to some large outlet.   The inlet should be 10 or 12

inches in diameter and have two rods placed vertically through drilled holes near the entrance to serve as a screen. Both inlet and outlet pipes should be vitrified sewer-pipes to withstand the effects of freezing, which common clay ware will not. A wooden box instead of a brick well will serve the same purpose, but must

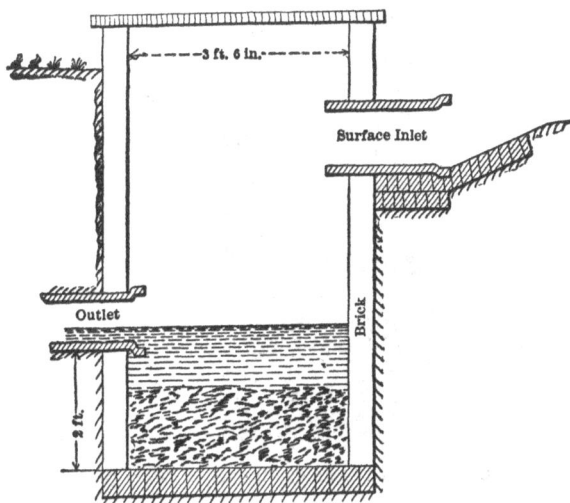

FIG. 34.—Catch-basin for Connecting Surface Drainage with Underdrains.

be renewed when the wood decays, and is obviously less permanent as an improvement. In some respects, however, the wood construction is more desirable. The inlet may then be a hole in the side of the box at the surface flow line, with iron rods fastened vertically across it.

A serviceable and easily constructed catch-basin may be made of the sections of 24-inch sewer-pipe, which should be set end to end in a vertical position and the

socket joints secured by cement mortar.   The bottom
section should be a straight pipe, the upper two sections should have T's, one of which should be used
for an outlet and the other for an inlet.   A 2-inch
plank cover fitted and dropped into the top socket fin-
nishes a neat and desirable catch-basin.   See Fig. 35.

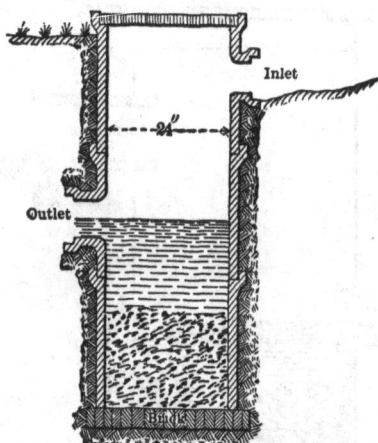

FIG. 35.—Catch-basin constructed of Sewer-pipe.

These catch-basins will require some personal attention at times.   Straws and other surface rubbish from
the surface ditch will gather against the inlet and must
be removed.   The mud which accumulates in the bottom of the basin should be removed from time to time.
It is, however, a useful and convenient accessory to
drainage work.

Drainage of Cellars and Residence Grounds.

Cellars which are excavated in clay or loam soils
usually become wet and require drainage.   A common

method of procedure is to construct a tile drain to carry the water away after it enters the cellar. Another is to cover the walls and bottom with a thick coat of cement mortar to prevent water from entering. If there is much soil water to contend with, it frequently bursts through the coating used in the latter method, and when the former is used, the cellar remains damp even when the drain removes the free water which percolates through the soil and finds its way to the low point in the bottom from which drainage is made.

The proper plan to follow is to prevent water from entering by means of a tile drain, which should be laid entirely around the building 4 or 5 feet distant from the walls and nearly if not quite as deep as the floor of the cellar. This drain should be of 4-inch tile laid on an even fall of 1 inch in 16 feet, and connect with a main which will carry the drainage safely away from the house. By this method the ground about the house as well as the cellar will be kept dry and wholesome, and is the best known plan for securing a dry cellar for a country house where natural drainage is deficient.

The underdrainage of the ground occupied by buildings and surrounding yards and gardens is often neglected on the supposition that these grounds have sufficient natural drainage, which is not always the case. The filtration of all surface water through the soil of the lawn, garden, and surrounding grounds in general, and its removal by the process of underdrainage, will add much to the ease with which walks may be constructed and maintained, and to the satisfactory growth of all useful and ornamental plants which contribute so largely to the beauty, value and healthfulness of a country residence.

The water supply is usually taken from a well located near the residence. While the supply of this well may have its source in a vein of clay, and be all that is desired in point of purity and coolness, it is subject to contamination by surface and soil water which percolates through the earth and finds lodgment in the well. Some deep underdrains laid about the well will prevent the pollution of the water from this source where the wells are located on level or undulating tracts of land.

Drainage of Fruit Orchards.

Much has been said and written upon the subject of growth and subsequent fruitage of orchard trees in various localities. That there are climatic and soil conditions beyond our present knowledge which have a bearing upon this industry cannot be denied. It has been observed that orchards of apple-trees set out upon land which had far from perfect drainage throve and produced excellent fruit for a series of years. Later, trees which were planted on an adjoining tract under apparently more favorable conditions failed to make satisfactory growth or to produce fruit of good quality in paying quantities.

There are examples in great number, however, which show the value of underdraining fruit orchards which are located upon close clay soils, or others which are too wet for the growth of cereal and garden crops. Fruit-trees will not grow well on wet land. The drains should be placed midway between the rows of trees and extend with the slope of the land to an intercepting main. (Fig. 36.)   These drains should be laid 4 feet deep if possible.   It will be found that while the larger part of the roots of fruit-trees are in the first 4 feet of soil, there are roots which extend vertically much deeper than this in search of moisture when it is lacking nearer the surface.   No stoppage of drains by roots of fruit-trees has been noted, but it should be observed that this is not the case with trees

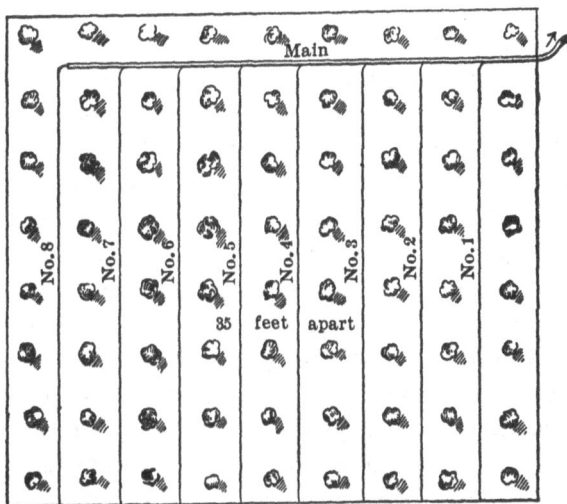

FIG. 36.—Orchard Drainage.

like the willow, water-elm, and others which are aquatic or water trees by nature, and whose roots have been known to seek out and clog tiles where growing 30 or more feet distant from the drain.   Drains which are dry during a large part of the year are rarely found to contain roots of trees of any variety, while those drains which have a constant stream of water running through them attract the roots of aquatic trees. The tiles should be laid with cement-covered joints where they come near such trees or the trees should be destroyed. The experience of fruit-growers along this line shows that the decay of growing trees may often be arrested by deep underdrainage. An eminent horticulturist says that a clay soil is not worth the taxes for apple culture if not underdrained. Every orchard planted on clay soil should have an underdrain between the rows of trees, and it should be 4 feet deep if possible. Dr. W. I. Chamberlain, of Ohio, in commenting on the drained and undrained portions of his own orchard, says: " I have now nearly finished picking and marketing the Red Astrachans on a row which, like all the other varieties, runs across both plats. The total yield per tree is fully 50 per cent greater on the tiled part, and in size, beauty, and evenness of shape there is more than that amount in favor of the tiled.'' With reference to the growth of the trees from the time of setting to maturity, he says that on the untiled land he lost 63 per cent of those planted, but on the tiled part of the orchard only 13 per cent perished. Those interested in fruit culture, and especially those contemplating the planting of large orchards, should look into this subject closely. If 50 per cent more trees can be brought to bearing from the first planting and the abundance and quality of the fruit materially increased by thorough underdrainage, it follows that both new and old orchards on clay land or black soil with clay subsoil should be drained. These are facts which do not stand alone, but can be emphasized by similar ones in other localities.

CHAPTER XIII.

ROAD DRAINAGE.

At this time good drainage is recognized among road-builders as a necessary part of the construction of roads, be they common earth roads or those improved by a metal covering. Much valuable information relating to the construction and maintenance of public roads has been collected and disseminated by the Office of Road Inquiry of the U. S. Department of Agriculture, and the bulletins issued from that office form a valuable compendium of road practice in this country. From these it is not difficult to learn that underdrainage either natural or artificial, as well as surface drainage, is held in high esteem by all who have experience in constructing roads over good agricultural soils. Drainage has to do with the durability and maintenance of a road after it is once constructed. A road may be constructed without proper attention to foundation and drainage, and, like a building erected under similar conditions, may appear well at first, but will soon show weakness, with threatening dissolution.

The economical and successful maintenance of a road involves the following principles:

1. The travelled surface should be so shaped and of such hardness that all storm water will flow off readily. Could the earth road be made so that this

condition would be secured, the question of how to obtain good roads in the country would be answered. Soil absorbs water easily, so that it is scarcely possible to make a surface that will shed it all, especially in climates where the surface is subject to periodic freezing, thawing, and rainfall. Roads which are excellent during the summer season when the road surface can be kept intact, lose all semblance to their summer and fall conditions during the winter and spring months, mainly because the surface is not water-proof.

2. Surface ditches at each side should be provided and so graded that storm water from the road surface will be removed quickly.

3. In order that the travelled surface or road crust may have a firm foundation at all seasons of the year, underdrains should be laid on one or both sides of the road-bed near its base, to prevent the saturation of the subsoil beneath the road.

These three elements are essential to every good road which is constructed on loam or clay lands. A thoroughly drained sub-grade must be provided if we expect to maintain a good surface, whether it be of earth or of some better material. The difference in the practice of the English road-builders Macadam and Telford gave rise to sharp discussions regarding the proper foundation for stone roads. The former claimed that the base should be formed of the larger size of broken stone which were used in making the road covering, while the latter held that the base should be formed of large flat stones, upon which the broken stone should be placed. With thorough sub-drainage there is no cause for such discussions. It is only where such drainage cannot be obtained that a base of large stones or timber must be added for a foundation support. While great progress has been made in the improvement of roads in many localities by covering with some available road metal such as stone, gravel, brick, and slag, the improved earth road is the base of all, and of necessity must be the road used by the mass of people in country districts; hence the principles of its construction and maintenance should be familiar to every one.

The amount of travel which passes over a road has a great deal to do with the completeness with which underdrainage effects it. Where roads are but little travelled, as in the case of farm roads and lanes used by a neighborhood, the simple drainage of land by one line of tile lengthwise through low places has proved sufficient without grading. If one track becomes rutted, there is abundant room to make another by its side, and not sufficient travel to require both. Where the travel is heavy, as it is over all leading roads to towns and railroad stations, the case is entirely different. When the entire width of the road becomes once cut into ruts, and in wet weather puddled on the surface by the continued passage of loaded teams, no water will pass through the soil to the drains, and without an embankment and side ditches the road will grow worse. The underdrains will keep a good base upon which to build a road, but they will not take water from a puddled surface. On all such roads experience has proven that we need the combined action of under and surface drainage, together with the continual oversight and care of the travelled surface. One fact, patent to every observer of an underdrained road, is enough to prove the necessity of surface drainage, and that is this: That when by any means a rut has been made and puddled by the continual action of the wheels, no water will pass through the soil at those places. The water must be evaporated or the puddle tapped by a surface ditch. The surface of a road must be sufficiently crowning to throw off all water possible, and then underdrains at the side will prevent the bottom from becoming saturated and soft. The surface should receive such care at times as is necessary to keep it crowning and to drain water from ruts that are sure to introduce themselves into all earth road surfaces. A line of tile at each side of the embankment gives the most perfect drainage. Let the lines be so located that teams cannot drive directly over them, for their value for taking surface water depends upon some part of the earth near the drain being left porous and free from the puddling effects of passing teams.

Deep side trenches are not desirable, though they may serve the purpose of lowering the water level sufficiently for road purposes, because they are difficult to keep graded so that the water will all flow from them, and because the mowing-machine cannot be conveniently used to keep the grass and weeds trimmed down, which is also a desirable part of road maintenance. Besides this, these side ditches are used for a winter sleigh road in latitudes where there is snowfall, and deep ditches are usually narrow and are ill suited for such use. Broad shallow side ditches graded so that storm water may flow away with tile drains at the inner edge to remove subsoil water form the best known plan for road drainage. This method of drainage is shown in Fig. 37, which represents a road surface

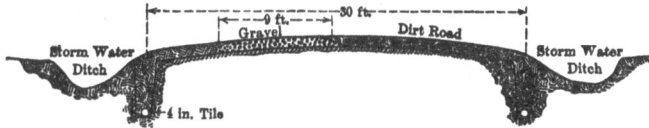

FIG. 37.—Proper Location for Surface Ditches and Tile Drains in Road Construction.

18 feet wide, one half of which is gravel surface and the other half earth surface. This is regarded as the most economical method of road improvement at low cost now in use where a hard road is desired. The earth track is the favorite one for use in dry weather and for light loads, while the gravel track is used for heavy loads and at times when the dirt track is wet. In this way the wear is shared by the two and both tracks easily kept in order.

Sub-drainage has lessened the cost of all hard roads by demonstrating that a less thickness of road covering or road metal is required than was formerly thought necessary. All experienced road-builders now emphasize the importance of complete underdrainage in constructing hard roads. With a firm foundation which underdrainage secures, it is only necessary to construct and maintain a good wearing surface. With good soil drainage, much of the height of large embankments may be reduced by lowering the line of saturation instead of raising the road surface by costly earth work where this is not required for the purpose of obtaining an economical grade.

Seepage Water on Roads.

It is not uncommon to encounter water veins or "spouty" places where excavation for road grades is made in grading hills. When these are not provided for, they present one of the most serious obstacles to the maintenance of the road surface. The remedy for this defect may in most cases be easily and effectively applied. Find where the water comes from and inter-

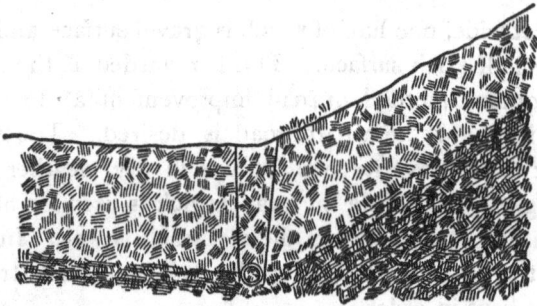

FIG. 38.—Tile Drain to Intercept Seepage-water.

cept it by means of a few tile drains before it reaches the base of the road-bed. Lay the drains in the soil where the water is found and extend the outlet line to the nearest available exit. The grade of the lines and workmanship in general should be as carefully looked after as in field drainage. Two hundred feet of 3-inch or 4-inch tile wisely used will frequently abolish a troublesome mud-hole on a hill road.

Surface Drainage of Hill Roads.

Roads on hills which have a grade of 3 per cent or more are frequently injured by storm water making a channel in the middle, or in the wheel tracks of the travelled road, which soon renders it unfit for use. To prevent this the surface of a dirt road should be made crowning as much as ten inches in a 20-foot roadway. The side ditches will suffer much by erosion and irregular washing unless the flow is controlled by occasional cutoffs by means of small cross-culverts, which will divert and discharge the water at the lower side of the road right-of-way without injury to either road or adjoining land. These cross-drains should be located at favorable points along the grade, and should consist of good sewer-pipe not less than 10 inches in diameter laid diagonally across the road track. With proper selection of the location and the adoption of the plans suited to the work, the top of the culvert may be placed 18 inches below the surface of the road and will constitute a durable improvement worth many times its cost. The joints of the pipe should be laid in good cement mortar, and should, if possible, have a grade of 1 foot in 20. These cross-drains being laid in solid, not filled earth, are not open to the objection urged against pipe culverts which are laid for waterways and covered with loose embankment.

Sewer-pipe, Culverts, and Cross-drains.

Experience with large earthen pipes for road culverts has demonstrated that they frequently fail, not by reason of lack of strength to resist the compression that they may be called upon to bear, but because of the jar or vibration which is communicated to them through the material with which they are covered. These failures are notable in locations where the pipes are bedded in and covered with gravelly, rocky, or other loose material which does not become thoroughly compacted. When a load pasess over them, the material conveys a vibration to the covered pipes which subjects them to a series of shocks which in time shatters the brittle and rigid material of which sewer-pipe is composed. When the pipes are full of frost or surrounded by ice, as they are at times in cold climates, the ware is especially subject to fracture by shocks. That this is the cause of failure rather than the superincumbent weight or compression is evidenced by the shattered condition of the sections, whereas if they failed by pressure the breakage would be in cracks parallel to the length of each piece. The double-strength sewer-pipe has given but little better results than the regular thickness for the obvious reasons already noted. Railroad companies have discarded sewer-pipe and are substituting cast-iron pipe for small culverts, since it was found that the tremor of embankments caused by passing trains destroyed the value of the sewer-pipe culvert.

Sewer-pipe may, however, be used for road cross-drains and culverts in good clay soil in which the joints can be firmly embedded and the same material thoroughly compacted about the entire length and with the top line of the pipe not less than 2 feet below the surface of the road. The well-known property of clay soil to form a bridge or crust surface under continuous travel does much to carry the load, while the elastic material about the pipe relieves it from injurious vibrations. The superiority of soil or sand over gravel, stone, or other loose material for bedding or covering culvert-pipe has been proven by experience.

outlets for water at every cross-stream or ditch where it was possible to discharge the water. After the drains were put in, a strip of brick pavement was laid close to one of the drains, leaving 24 feet width of dirt road for summer use. This dirt was repeatedly rolled with a heavy roller until the upper foot or 2 feet of the crust of the road-bed became hard and solid. (See Fig. 39.)

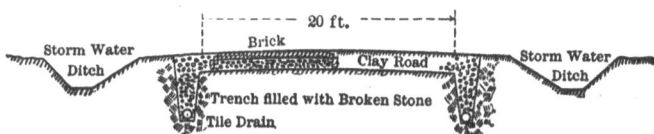

FIG. 39.—Combined Brick and Clay Road as constructed in Cuyahoga Co., Ohio.

Our work on that road has demonstrated that heavy rolling of a road which has been properly drained will form a crust or roof, so that water cannot stay on the road, but must flow at once into the drain-pipe and disappear; and in case of storm water too rapid for the pipe-trench to catch and carry off, the water flowed over the pipe-trench into the storm-ditches, which never fail to carry off all the water that comes. Since the road was finished there has been no break, no settlement, no stoppage of water, no ruts, no mud, and travel on the road has doubled many times, thus showing the popularity of a hard, even roadway for winter travel as well as summer.

" The method of holding the brick in place alongside of the dirt road was devised by me, and consists of three courses of brick standing endwise, the first course flush with the top of the pavement, the second breaking joints and dropping 2 inches lower, and the third 2 inches lower still, forming a stairwise bond for the brick-work in such a manner that a heavily loaded wagon cannot catch and tear up the brick pavement. If a wheel runs off the pavement it strikes the second course of curbing brick and runs along on that; but it is almost impossible for a wheel to cut through the broken stone filling which surrounds the curbing courses and protects them from the wear of heavily loaded wagons."

CHAPTER XIV.

DRAINAGE DISTRICTS.

A DRAINAGE district is an organization of the owners of land for the purpose of constructing and maintaining adequate drainage outlets for individual use in which the expense of the work is shared by each in proportion to the benefits derived.

The boundary of a district should in all cases, if possible, include an entire water-shed, so that when an outlet is secured and drainage provided for, it will be complete in itself as respects its drainage. The formation and management of districts is provided for in many States by laws which direct in detail the steps which should be taken, and give methods of procedure which must be followed closely in order to make the proceedings legal.

The principles now fairly well recognized in the execution of all public land drainage where individual property is affected and which must bear the expense of the improvement may be stated as follows:

The plan for the work should be complete and provide each landowner an adequate outlet which he may use without being accountable to neighboring property. Each property should be assessed for the first cost and subsequent maintenance of the work in proportion to the benefit it will derive from the same, taking into account the drainage advantages which each property possesses by nature.

The work should be undertaken only after it has been shown that the benefits which will accrue to the properties concerned will be greater than the expense of the improvement, and should be executed with due regard to economy and permanency.

Plans.

The natural boundary of the area should be determined with accuracy. Where the drainage divide is not apparent from observation, levels should be taken to determine the dividing line between water-sheds, for upon this point will often depend the legality of assessments upon certain tracts for the cost. There are table-lands or levels where there is no little difficulty in determining the natural divide. In fact some tracts may be drained by artificial methods as easily in one direction as in another, but the natural drainage of a tract as held by the courts is the direction in which the water will flow when the entire surface of the land becomes covered with water. The belt of land which first becomes dry when drainage takes place under such conditions is the natural divide, and should be regarded as the boundary line of the area to be included in a district. A map should be made showing the acreage and ownership of each farm, and upon this the boundary line of the district should be distinctly marked. The natural outlet for the district may be obvious, but its size insufficient, in which case it should be improved. The size and best dimensions for the

TWENTY-INCH DRAIN-TILE DISTRIBUTED FOR LAYING THROUGH A SWAMP. *(To face page 200)*

work should be determined by the methods outlined in previous chapters. The topography should be found in detail sufficiently to furnish data for the necessary computations for defining the dimensions that will be required for the main outlet and for such branches as may be needed to provide for the drainage of individual lands. The provision of outlets for the property of owners located at some distance from the main channel or drain will incidentally require the construction of drains across farms which will thereby be much more completely drained at the general expense than other tracts for which a mere outlet is constructed to the property line. In such cases there is a difference in the amount of benefit conferred by the work which must be considered in apportioning the cost to the several owners.

The system may consist of open ditches supplemented by large tile drains, or of either separately, according to the area to be drained and the requirements of the land which will be affected. Where it is desirable to use large tiles for mains, and to lay them in the line of natural drainage courses, it is always wise to provide overflow ditches which will relieve the surface at times of unusually heavy rainfall and at the same time retain the benefits which are peculiar to under-drains. (Fig. 40.) These surface drains should be shallow and broad and so graded that while all water may be removed from the surface by the open ditch, the flood water will pass off in sufficient quantities to permit the tile drain to complete the work perfectly. The surface drains may be called into action only occasionally, but at certain times save crops worth many times the cost and care of such drains. Experience has demonstrated that in well-watered sections tile drains which are intended to provide outlets for large tracts cannot be made large enough without too great expense to take care of exceptionally heavy rainfall with-

FIG. 40.—Tile Drain with Relief Surface Ditch.

out the aid of surface drains. Another reason for the necessity of this provision is that many soils which under ordinary conditions will respond readily to the action of tile drains in removing water from the surface through the soil will not permit a sufficiently large quantity of water to pass to the drains in the short time required. If the ditches are left broad and shallow they will cause but little inconvenience and may be utilized for cultivated field crops or for grass.

An estimate of the cost of the entire proposed work, including construction, legal and administration expense, should be made and apportioned to the several property owners and interests concerned in the work.

The Theory of Classification of Lands.

The first step to be taken in determining the apportionment of cost which each tract of land should bear is to fix upon some scale of marking which shall numerically express the benefits to each tract of land, and which may be used in making a just and equitable

THIRTY-INCH DRAIN-TILE DISTRIBUTED ALONG THE LINE OF A SURVEYED DRAIN.

(To face page 202.)

distribution of the entire expense connected with the construction and maintenance of the projected work.

This is one of the most delicate matters involved in cooperative drainage, since there are many different opinions among landowners and others concerning the proper distribution of cost, the judgment of some interested parties being frequently biased by personal considerations. It may be asserted with truth that the classification of land for this purpose, even when made with the utmost care and good judgment, will be open to just criticism from some point of view.

Each of the following principles should have a value in the classification of land in a drainage district:

Each landowner is entitled to such natural drainage as his land possesses by right of ownership. If his land is so situated that he can thoroughly drain it into channels provided by nature without crossing his neighbor's land in the construction of drains he should not be required to assist in carrying out a plan of cooperative drainage, except on the ground that the proposed work will promote the public health or enhance the general value of property in the community.

Each landowner may drain his land as he chooses provided he does it within the boundary of his own possessions, into an outlet channel provided by nature.

A tract of land which is wet and practically useless for agricultural purposes should be assessed proportionally higher if reclaimed by the drainage system than that which has better natural drainage.

A tract which lies distant from a natural outlet should be taxed higher than one lying near, provided both receive the same drainage advantages. This obtains on the theory that a tract near natural drainage in a state of nature bears a higher value by reason of this fact than one which is distant and whose lack of natural drainage is recognized as a cause of diminished value.

In case a public drain passes through a farm for the purpose of giving drainage privileges to another farm, and in so doing incidentally diminishes the individual expense which will be required to complete the detail drainage of the first farm, it, the first farm, should be assessed higher proportionally than the second, on the theory that private drainage has been done on the first farm at general expense.

Classification for Assessment Purposes.

A classification map should be prepared which shows the name of the owner of each separate piece of land, the location and the kind of drains proposed and the estimated cost of each of the mains, and the total estimated cost.

With this map in hand, the party or parties to whom is intrusted the classification from which the assessment roll is to be prepared should make a personal examination of each tract or farm included in the district, noting carefully the natural condition and location of the land and the comparative benefit which each tract will receive from the proposed work. Use the number 100 to represent the classification of the tract, be it large or small, which will receive the greatest benefit considering the principles which have been heretofore stated. Compare other tracts with this and rate each on the per cent system, 40-60-70, etc., placing the classification number upon the respective tracts. This classification may be reviewed and amended until it is agreed that these numbers represent the proper ratios of benefit which the several tracts or farms will receive by reason of the construction of the drainage system.

In case the public highways, the township or county are benefited as organizations, and are legally subject to assessment for the same, they cannot be classified upon the same basis as agricultural lands, but a certain per cent of the entire cost should be charged up to each and deducted from the sum total, after which the remainder should be spread over the district as per classification.

All of this work is a matter of judgment, and upon it will depend the equitable and proper distribution of the cost. It cannot be done too carefully, and should be several times reviewed before it is finally approved.

To find the part of the total cost that each owner should pay, proceed as follows:

1. Deduct from the total estimated cost of the work the several amounts assessed against highways, townships, corporations, etc., to obtain the amount that is to be charged to the classified lands.

2. Multiply the number of acres in each tract by its classification mark.

3. Find the sum of these several products.

4. Divide the amount to be apportioned to the several tracts by the sum of the several products and multiply each product by this quotient to obtain the amount that should be assessed against each tract as shown upon the roll. These several amounts when added together should equal the total estimated cost.

The assessment roll and classification map here given (Fig. 41) will serve to illustrate the methods

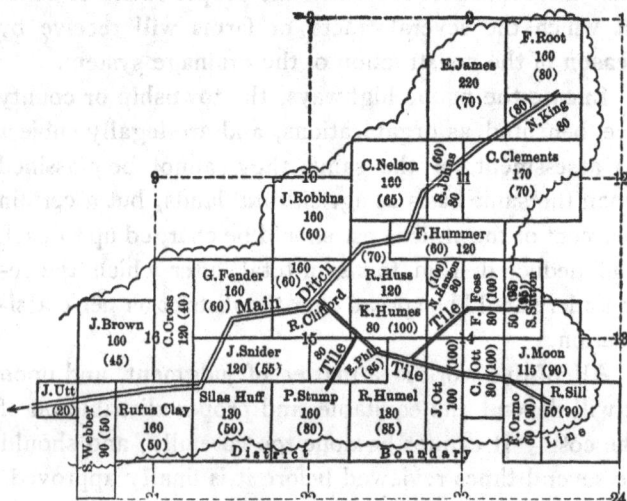

Fig. 41.—Classification Map of a District containing 3525 Acres.
(Classification figures in parenthesis.)

described and in a measure the principles of land classification for district purposes.   In addition to the columns shown in the roll here given the description as to section, etc., of each owner's land should immediately follow the name.

There are always many questions peculiar to each case coming up in this kind of drainage work which must be met and disposed of in the wisest manner possible.   The men who have charge of the work as commissioners should be conversant with land-drainage matters in general, and with the proposed work in particular, in order to be able to exercise good judgment regarding the effect that the execution of the contemplated work will have, and be ready to give proper consideration to suggestions and opinions which may be advanced by interested landowners.

ASSESSMENT ROLL OF CRESCO DRAINAGE DISTRICT.

Total estimated cost $6280.00
Amount assessed against township highways 628.00

Balance to be assessed against farm lands $5652.00

Names of Owners.	Acres.	Classifi- cation.	Product.		Assess- ment.
J Brown	160	45	7,200	.02301	$165.70
S. Webber	90	50	4,500		103.57
Rufus Clay	160	50	8,000		184.10
C Cross.............	120	40	4,800		110.46
Silas Huff	130	50	6,500		149.59
J. Snider	120	55	6,600		151.89
G. Fender	160	60	9,600		220.93
R Clifford	160	60	9 600		220.93
J. Robbins..........	160	60	9,600		220.93
E. Philo	80	85	6,800		156.48
R. Stump	160	80	12,800		294.56
R. Humel	180	85	15,300		352.09
J. Ott	100	100	10,000		230.13
C. Ott.	80	100	8,000		184.10
F Orno	60	90	5,400		124.27
J Moon.............	115	90	10,350		238.18
R Sill	50	90	4,500		103.56
S. Sutton	50	95	4,750		109.31
F. Foss	80	100	8,000		184.10
N. Hanson..........	80	100	8,000		184.10
K Humes..........	80	100	8,000		184.10
R Huss	120	70	8,400		193.32
F. Hummer	120	60	7,200		165.70
J Utt	40	20	800		18.42
C Nelson	160	65	10,400		239.33
S. Johns	80	60	4,800		110.47
C. Clements	170	70	11,900		273.85
N. King	80	70	5,600		128.88
F. Root.............	160	80	12,800		294.56
E. James...........	220	70	15,400		354.39
	3525		245,600		$5652.00
Public highways	....	....	....	....	628.00
					$6280 00

$$\frac{5652}{245,600} = .02301.$$

NOTE —The column headed "Product" gives an equivalent number of acres classified at 1 per cent.

The number .02301 represents the amount of assessment on one acre classified at 1 per cent.

Mutual Cooperative Drainage.

The same principles of work may be applied to small drainage schemes as those which are employed in drainage districts in which it is not necessary to work under legal restrictions. In other words, it may be a partnership plan, mutually arranged and agreed upon, from which will result the same advantages without incurring the delay and expense involved in operating under the statute.

A plan, estimate, and assessment may be submitted by the engineer, made out by drainage-district methods and submitted to the parties concerned for their concurrence. If it is acceded to, nothing stands in the way of entering into an agreement and executing the work according to the accepted plans, in which case the necessary expenses will be reduced to a minimum. This is a commendable method when interested parties can agree.

CHAPTER XV.

ESTIMATES OF COST.

An estimate of the cost of a proposed drainage work is expected of the engineer, and is important before the execution of the work begins. In the case of the drainage of a field or farm, a preliminary estimate may be made from an inspection of the land based upon a knowledge of what the drainage of other tracts of land similar in requirements has cost. No correct estimate can be made, however, until the lines have been located and measured and the sizes of tile to be used decided upon.

The cost of material and labor in the locality where the work is to be done must be ascertained and tabulated. The following is a schedule of the items that should be considered in estimates for farm drainage:

1. Total number of each size of tile required.

2. Cost of tile at factory.

3. Cost of freight to nearest railroad station, if shipped.

4. Hauling from factory or station and distributing on the drain lines.

5. Digging ditches and laying tile.

6. Filling ditches.

7. Laying out and superintending.

Schedule for Making Estimates.

The following schedule, subject to amendment as prices vary, may be used in making estimates. The weight of the individual tiles varies somewhat with different factories, as does also the length and diameter. The joints from some factories are 12½ inches long, while from others they are even 12 inches. The prices here listed are, at this date (1902), considered fair average prices at the factory. The joints of tile 12 inches in diameter and above should be 2 feet long for convenience in handling and for strength.

PRICES AND WEIGHTS OF DRAIN-TILES.

(NOTE.—Prices vary in different localities.)

Size in Inches.	Price per 1000 Feet.	Weight per Foot Lbs.	Average Car Load, Feet.*
3	$10.00	4½	6000
4	15.00	6½	4000
5	21.00	9	3000
6	27.00	11½	2200
7	36.00	14	2000
8	45.00	18	1250
9	54.00	21	1000
10	60.00	25	850
12	90.00	33	750
14	135.00	43	600
15	150.00	55	500
16	165.00	62	450
18	240.00	80	350

* If large and small sizes are loaded in the same car, the freight cost will be lessened on a field list of tile.

Hauling and distributing tile can be figured by the cost per ton, using as a basis the number of loads which can be hauled per day, when the road is good, by one man and team.

With wages of team and driver at $3.00 per day, two men, each with a team, working together so that one can assist the other in loading and unloading, hauling and distributing tile will cost approximately as follows:

Hauling one mile...... $0.55 per ton
     "   two miles 70   "   "
     "   three miles....   1.00   "   "
     "   four miles.....   1.25   "   "
     "   five miles......   1.40   "   "

The tile should be strung along the lines staked out for work if it is to be done soon; if not, they should be placed in neat piles of 25 each, at regular intervals of 25 feet, near the line of each drain. A sketch or diagram of the location and sizes should be given to the one in charge of the distribution so that this part of the work may be done correctly.

The difficulties to be encountered, such as the condition of the roads, the fields, and inconveniences of loading and unloading, must be taken into account, which will often materially change the foregoing figures that should be used in estimating the cost of the delivery of tile upon the ground.

Digging the ditches and laying the tile are so frequently contracted for and done by the rod or 100 feet, that a price is pretty well established for drains in soils which are easily worked with the spade and shovel, and which is here stated as a basis upon which to work.

Where the land is stony, filled with roots or is so hard that it must be picked or loosened with mattocks, the cost will be much increased and must be figured with a liberal margin in order to cover unforeseen contingencies.

COST OF DIGGING DITCH AND LAYING TILE TO GRADE.

3″ to 6″ tile	2 to 3 ft. deep,	$1.20 per 100 ft.
" " " "	3 to 4 " "	1.50 " " "
" " " "	4 to 5 " "	2.10 " " "
7″ and 8″ tile	3 to 4 " "	1.80 " " "
" " " "	4½ " "	2.40 " " "
" " " "	5 " "	3.50 " " "
9″ to 12″ "	4 " "	3.00 " " "
" " " "	5 " "	3.60 " " "

Filling the ditches in cultivated land can be done with a team and plough for 10 cents per 100 feet. In meadows where care must be taken to leave the surface uninjured, for 15 cents per 100 feet. Where the filling must all be done by hand work it will cost about 25 cents per 100 feet.

Laying out and superintending are worth about 5 per cent of the total estimated cost where nothing is required but location and levelling lines and final inspection of the drains before they are covered.

TABLE SHOWING THE NUMBER OF CUBIC YARDS OF EARTH IN ONE ROD OF DITCH OF VARIOUS DIMENSIONS.

(Excavation for Tile Drains.)

Depth. Inches.	Mean Width.											
	7 In.	8 In	9 In.	10 In.	11 In.	12 In	13 In.	14 In.	15 In.	16 In	17 In.	18 In.
30	0.89	1.02	1.15	1.27	1.40	1.53	1.65	1.78	1.91	2.04	2.16	2.29
33	0.98	1.12	1.26	1.40	1.54	1.68	1.82	1.94	2.10	2.24	2.38	2.52
36	1.07	1.22	1.38	1.53	1.68	1.83	1.98	2.14	2.29	2.24	2.60	2.75
39	1.16	1.32	1.49	1.65	1.82	1.98	2.15	2.32	2.48	2.65	2.81	2.98
42	1.25	1.42	1.60	1.78	1.96	2.14	2.32	2.49	2.67	2.85	3.03	3.21
45	1.34	1.53	1.72	1.91	2.10	2.29	2.48	2.67	2.86	3.05	3.24	3.44
48	1.43	1.63	1.83	2.04	2.24	2.44	2.65	2.85	3.05	3.26	3.46	3.67
51	1.52	1.73	1.95	2.16	2.38	2.60	2.81	3.03	3.25	3.46	3.68	3.90
54	1.60	1.83	2.06	2.29	2.52	2.75	2.98	3.20	3.44	3.66	3.89	4.12
57	1.69	1.94	2.18	2.42	2.66	2.90	3.14	3.38	3.63	3.87	4.11	4.35
60	1.78	2.04	2.29	2.54	2.80	3.05	3.31	3.56	3.82	4.07	4.33	4.58

Total Estimate.

By going over the foregoing items in detail and adapting the prices to the locality, a correct estimate may be summed up for the cost of the work on the entire field, farm, or tract, which should be charged to the number of acres which will be benefited by the proposed improvement.

From what has been said in previous chapters upon the frequency and depth of drains, it is readily seen that there will be a wide difference in the cost of work according to locality and kind of soil operated upon. The cost per acre for the entire tract improved, rather than the cost of individual drains, should be sought.

Profit on the Investment.

How much will the land be benefited? How much will the production be increased? What will be the saving in labor of cultivation and general management of the land? These things and many more enter into the account to be figured upon, all of which will have a bearing upon the profits, not merely one year, but of all future years.

The non-resident owner looks at it from a strictly investment standpoint. The improvement will pay him a certain desired per cent on the outlay. If it figures out satisfactorily, the money is placed *in the ground* instead of upon it. Hence the necessity of being able to strike a proper and correct balance between estimated cost and estimated profit.

It does not come within the province of this chapter to show what the actual profit of such work will be, but to outline the ground to be gone over in arriving at the cost of drainage in any locality in order that the figures may tell their own story in a comprehensive way. It may be remarked that some land will pay a large profit on the cost of drainage, while other land is not worth it.

Cost of Work under Drainage District Organization.

In case of drainage work which must be done under the provisions of a State drainage law, great care should be taken by the engineer and others who have direction of the proceedings to act strictly in accordance with the directions of the statute in every particular. The plans and estimates form the basis of all subsequent work and should be most carefully prepared and considered.

The cost of each drain, whether open or of large tile, which is used for an outlet should be estimated separately, and be so scheduled, for the reason that in the classification and assessment of lands according to benefit conferred the cost of separate drains should be taken into account. The cost of the various items for which the law makes the organization responsible should be closely canvassed.

The cost of engineering, and of administration, which includes legal expenses, fees of commissioners, and superintendence, should be added to the cost of the execution of the work. This, of course, will vary with the size of the district and the legal difficulties that may arise, but is usually not less than 15 per cent of the cost of construction.

Offsetting the estimated expense of the work should

be placed the estimated benefits which will result and which the engineer should be prepared to make. It is needless to say that in order to do this he should be conversant with the value and management of lands. A clear and logical statement of cost and benefits involved in a proposed improvement, based upon facts which will commend themselves in a forceful way to the court, and to landowners, will be most valuable, and if properly presented will do much toward avoiding legal controversies which arise by reason of a misunderstanding of the facts that should be considered in the case.

CHAPTER XVI.

BENEFITS AND PROFITS OF DRAINAGE.

THE benefits and profits accruing from the drainage of land should be set forth comprehensively when the Court or Board of Commissioners inquires for them during investigation of any drainage project brought before them for consideration. "The cost of the improvement must not be greater than the benefit," is the brief rule set forth in the drainage laws, and is a commercial maxim which is recognized in general business. "Will it pay?" is imprinted upon the initial page of every business undertaking, not in pleasure and satisfaction, but in dollars and cents.

Some of the benefits and resulting profits of the work under consideration may here be enumerated as suggestive rather than carried out in detail.

Firmness and Fineness of Soil.—One of the changes produced by relieving the ground of surplus water at once noticeable is its increased firmness. The excess of water recedes from the surface and takes its place lower in the soil, soon leaving a firm surface which can be passed over by teams or live stock without injuring the texture or destroying the smoothness of the surface. Not only does this facilitate the culture and management of land, but the travelling public get benefit and profit by reason of better roads which result from the improvement. The fineness of the soil is increased by the percolation of water from the surface downward through the soil, which permits air and frost to do their work more effectually in disintegrating the particles of soil, reducing their size and increasing the capacity of the soil for moisture. It has been shown in a previous chapter that the per cent of moisture held in the soil after the surplus has been drained off increases in proportion to the fineness of the particles composing it.

Permits Earlier and More Timely Cultivation.— One and sometimes two weeks' time in the spring are gained to land and roads by good underdrainage. The water from rains and thawing ice passes down through the soil, admitting warm air and fertilizing rains to such an extent that the surface is well prepared for the crops requiring early planting much sooner than wet soils. This is of great advantage not only to the cultivation on the season's work, but often makes the difference between an excellent and profitable crop and an indifferent one.

Produces Aeration of Soil.— A certain degree of soil ventilation has been found necessay to bring heavy lands to their highest state of productiveness. When the level of the ground water is lowered, the roots of plants penetrate more deeply into the soil, and as they die and decay leave a network of channels extending to the surface, through which air circulates, forming ventilating shafts, as it were. The interspaces of soil becoming relieved of water are also filled with air, which carries with it any fertilizing gases it contains and furnishes oxygen to the roots of plants and for the support of soil bacteria, which are now recognized as playing such an important part in converting soil humus into nitrates. These elements of fertility are absorbed by the soil and held in readiness for the use of plant-roots. The changes wrought by the passage of water through the soil and consequent circulation of air are continuous processes and play no unimportant part in keeping up the productiveness of soils. In other words, well-drained soils do not become exhausted as soon as undrained ones, even under the most ordinary treatment.

Drainage Increases the Temperature of the Soil.— A soil cannot become warm until the water upon its surface is evaporated or thoroughly warmed by the sun. The cause is easily explained. A large amount of heat is used up in evaporating the excess of water. Prof. F. H. King in his work, "The Soil," gives deductions from personal experiments which are valuable in showing the effect of drainage upon soil temperature. He says: "While 100 heat units will raise the temperature of one pound of water through 100° F., it is necessary to use 966.6 heat units to evaporate one pound of water from the soil; but this if withdrawn directly from the cubic foot of saturated clay would lower its temperature about 10.3° F. It must be evident, therefore, that to allow the surplus water to drain away from a field rapidly, rather than to hold it there until it has time to evaporate, must greatly favor the warming of the soil."

He cites the following observations showing the differences of temperature in the surface inch of well-drained sandy loam and an undrained black marsh soil, both of them bare and level.

Date.	Temp. of Air	Temp of Drained Soil.	Temp. of Undrained Soil.	Differences.
April 24............	60.5° F.	66.50° F.	54.00° F.	12.50° F.
" 25	64.0° F.	70.00° F.	58.00° F.	12.00° F.
" 26...........	45.0° F.	50.00° F.	44.00° F.	6 00° F.
" 27...........	53.0° F.	55.00° F.	50.75° F.	4.25° F.

The above observations are upon unlike soils, but the range of differences in temperature between a drained and undrained soil from which the surface-water has been removed may be taken at from 5 to 10 degrees.

Drainage Prevents a Large Waste of Fertility by Surface Washing.—The object of all drainage should be to remove all surplus water through the soil, not over it, thereby preventing loss by washing away the fine soil particles which constitute its richest part. The drained soil acts as a filter to arrest all the fertility which may be held in suspension by the water to be removed.

Increases the Depth of Soil.—A drained soil becomes renovated—opened up, so to speak, to the full depth to which it is drained. An additional field is opened up for the use of plant-roots, giving them a larger range from which to obtain both food and moisture.

A Drained Soil Resists Drought.—With reference to the value of drainage for enabling soils to resist the inroads of drought, experience confirms what we might expect would be true from the general effects which drainage has upon the soil. The additional fineness of soil produced which makes it capable of retaining a greater amount of capillary water, the greater depth of soil from which plants may draw nutriment, the circulation through the soil of air from which the moisture is condensed in its cooler recesses, all contribute to the power of a drained soil to resist the effects of a protracted dry spell during the growing season.

The experience of farmers who have drained clay or alluvial soils sustains these conclusions as to value of underdrainage in time of drought.

Benefits from Open Channels.—Their value consists in making more complete drainage possible. Whenever the value of complete underdrainage is established, the value of open ditches is established if they become a necessity for properly carrying out the work. In the absence of natural receiving channels, artificial ones must be constructed, or old natural channels improved so that drainage may be completed by the owners of individual tracts of land. Open ditches often serve as the outlet of the soil drainage of towns and villages, or for the effluent of sewage after it has been treated or filtered in such a manner and so effectively that it will be harmless. In such cases assessment should be made on the corporation for its share of the cost of the ditch. Such an assessment cannot be made upon the same basis as upon agricultural lands, but upon the valuation of the property of the corporation compared with the value of a similar area of farm property.

These ditches are of marked benefit to the highways of the tract which they drain in giving outlets for road drains and in the work incidentally effected by the gen-

eral system.   The highway as such should be assessed
for these benefits, which sum should be paid out of the
general road fund.

Profits of Drainage.   What are They?—On the
farm they are:

1. Proceeds from waste land which before drainage
produced nothing—a clear gain of land which is worth
as much for productive purposes as the balance of the
farm.

2. Doubling the annual production of many other
acres without increasing the cost of cultivation.

3. Diminishing the expense of management by
reason of doing away with broken-up fields and unnecessary open ditches.

All of these represent a money value to the farm
which can be readily estimated if analyzed and placed
in order with a value attached to each item.

Regarding the improvement of larger tracts which
have no productive value until drained the expense
account may be outlined as follows:

1. First cost of land.

2. Cost of draining to fit it for production.

3. Interest on purchase price until it is prepared
to produce.

4. Taxes until it can produce.

5. Clearing, if timbered.

6. Houses, barns, fences, etc., necessary to fit it up
into productive farms.

The sum of these expenses will represent the cost
of the land ready for producing a crop.   If the land is
fertile there can be no question regarding the value of
the investment unless the ultimate cost runs higher

than other land which is improved in a similar manner, which is rarely the case.   The improvement of land of this character is susceptible of being figured out with reasonable accuracy beforehand.   Two items in the above schedule should receive particular attention, namely, the cost and efficiency of the proposed drainage, and the value of the soil for productive purposes when drained.

What will it cost to reclaim it and what will it produce after it is reclaimed and improved ? are the points to be investigated by the engineer and purchaser.

Conclusion.

It has been the aim of the writer of this book to place within the reach of the engineer, the thorough agriculturist, and careful buyer of unimproved land such information and practical details of work as will enable them to engage in the work of land drainage intelligently and successfully.   All has not been said which might be said upon this important subject. Doubtless many details have been omitted which it would have been well to insert.   Be that as it may, the book is not loaded up with useless matter nor with untried theories, but is practical and easily understood.

INDEX.

———

TABLES.

ILLUSTRATIONS.